TRIANGLES AND TRIBULATIONS

Acting with Technology

Bonnie Nardi, Victor Kaptelinin, and Kirsten Foot, editors

Tracing Genres through Organizations: A Sociocultural Approach to Information Design, Clay Spinuzzi, 2003

Activity-Centered Design: An Ecological Approach to Designing Smart Tools and Usable Systems, Geri Gay and Helene Hembrooke, 2004

The Semiotic Engineering of Human Computer Interaction, Clarisse Sieckenius de Souza, 2005

Group Cognition: Computer Support for Building Collaborative Knowledge, Gerry Stahl, 2006

Acting with Technology: Activity Theory and Interaction Design, Victor Kaptelinin and Bonnie A. Nardi, 2006

Web Campaigning, Kirsten A. Foot and Steven M. Schneider, 2006

Scientific Collaboration on the Internet, Gary M. Olson, Ann Zimmerman, and Nathan Bos, editors, 2008

Acting with Technology: Activity Theory and Interaction Design, Victor Kaptelinin and Bonnie A. Nardi, 2009

Digitally-Enabled Social Change: Online and Offline Activism in the Age of the Internet, Jennifer Earl and Katrina Kimport, 2011

Invisible Users: Youth in the Internet Cafés of Urban Ghana, Jenna Burrell, 2012

Venture Labor: Work and the Burden of Risk in Innovative Industries, Gina Neff, 2012

Car Crashes without Cars: Lessons about Simulation Technology and Organizational Change from Automotive Design, Paul M. Leonardi, 2012

Coding Places: Software Practice in a South American City, Yuri Takhteyev, 2012

Technology Choices: Why Occupations Differ in Their Embrace of New Technology, Diane E. Bailey and Paul M. Leonardi, 2015

Shifting Practices: A Reflective Inquiry into Technology, Practice, and Innovation, Giovan Francesco Lanzara, 2016

Heteromation, and Other Stories of Computing and Capitalism, Hamid R. Ekboa and Bonnie A. Nardi

The Good Drone: How Social Movements Democratize Surveillance, Austin Choi-Fitzpatrick, 2020

Virtually Amish: Preserving Community at the Internet's Margins, Lindsay Ems, 2022

Data Rules: Reinventing the Market Economy, Cristina Alaimo and Jannis Kallinikos, 2024

Triangles and Tribulations: Translations, Betrayals, and the Making of Cultural-Historical Activity Theory, Clay Spinuzzi, 2025

TRIANGLES AND TRIBULATIONS

TRANSLATIONS, BETRAYALS, AND THE MAKING OF CULTURAL-HISTORICAL ACTIVITY THEORY

CLAY SPINUZZI

The MIT Press
Cambridge, Massachusetts
London, England

The MIT Press
Massachusetts Institute of Technology
77 Massachusetts Avenue, Cambridge, MA 02139
mitpress.mit.edu

The MIT Press would like to thank the anonymous peer reviewers who provided comments on drafts of this book. The generous work of academic experts is essential for establishing the authority and quality of our publications. We acknowledge with gratitude the contributions of these otherwise uncredited readers.

This book was set in Stone Serif and Stone Sans by Westchester Publishing Services. Printed and bound in the United States of America.

Library of Congress Cataloging-in-Publication Data

Names: Spinuzzi, Clay, author.
Title: Triangles and tribulations : translations, betrayals, and the making of cultural-historical activity theory / Clay Spinuzzi.
Description: Cambridge : The MIT Press, 2025. | Series: Acting with technology | Includes bibliographical references and index.
Identifiers: LCCN 2024030351 (print) | LCCN 2024030352 (ebook) | ISBN 9780262552172 (paperback) | ISBN 9780262382779 (pdf) | ISBN 9780262382786 (epub)
Subjects: LCSH: Organizational sociology. | Qualitative research—Methodology. | Information technology.
Classification: LCC HM711 .S667 2025 (print) | LCC HM711 (ebook) | DDC 302.3/501—dc23/eng/20241212
LC record available at https://lccn.loc.gov/2024030351
LC ebook record available at https://lccn.loc.gov/2024030352

10 9 8 7 6 5 4 3 2 1

EU product safety and compliance information contact is: mitp-eu-gpsr@mit.edu

CONTENTS

ACKNOWLEDGMENTS

This book draws together strands from over twenty-five years working with activity theory and related sociocultural theories. In that quarter century-plus, I have benefited enormously from working with people in a variety of ways.

My first thanks go to the series editors, Bonnie Nardi, Victor Kaptelinin, and Kirsten Foot. Bonnie was especially helpful in reading iteration after iteration of this book, helping me to get not only the arguments but the style right. The anonymous reviewers also contributed greatly, providing not just criticism and pushback but also constructive feedback and citations.

Thanks also to two scholars who agreed to review parts of this manuscript in development: David R. Russell and David Guile.They were very generous with their time, and both gave me both encouragement and much-needed criticism. The book is far improved because of their contributions. Needless to say, any remaining flaws and shortcomings are despite their efforts.

I've also learned a lot from my generous and knowledgeable colleagues and students at the Activity Theory Summer School (ATSS), held in 2019 and 2021–2024—both through their presentations and in our conversations afterwards. My colleagues at ATSS include Yrjö Engeström, Annalisa Sannino, Stan Karanasios, David Allen, and Kristina (Kicki) Johansson. I'm honored to be involved in ATSS as well as at the Centre for Activity Theory at Högskalan Väst.

In researching the roots of activity theory in the work of Vygotsky, Luria, and Leontiev, I benefited greatly from communications with Michael Cole, Dmitry Leontiev, Paul Prior, and Anton Yasnitsky.

At the University of Texas at Austin, I have benefited from conversations about theory and methodology at all levels. At the Ph.D. level, students in my seminars on activity theory, actor-network theory, and qualitative research methodology have challenged me to think differently about these topics. At the MA level, working professionals in the human dimensions of organizations program have engaged with me in sophisticated discussions about qualitative research and persuasion. And at the undergraduate level, students in the class Designing Text Ecologies have helped me to think through the nuts and bolts of CHAT-based qualitative case studies.

My family have been very kind and patient as I regaled them with eighty-year-old gossip about Soviet psychologists. Thanks to them for their constant support.

Finally, thanks again to Yrjö Engeström, whose works were introduced to me early in my Ph.D. program by David R. Russell. For the last twenty-five-plus years, I have read, struggled with, and sometimes pushed back on these works, but I've always found them to be insightful and inspiring.

1 INTRODUCTION

For the last twenty-five years or so, I've conducted qualitative case studies in a lot of workplaces, examining how people use texts and technologies to communicate, coordinate, and collaborate at work. And in doing so, I've drawn a lot of triangles.

Why triangles? It's an accident of history.

WORKPLACE WRITING STUDIES IN SEARCH OF A FRAMEWORK

As a young Ph.D. student, I entered the field of writing studies, focusing on rhetoric and professional communication—that is, how people communicate and persuade in their professions. Specifically, I was interested in how to support users as their work was being transformed by newly available digital computers and software. In retrospect, I was trying to answer the following questions: *How do people communicate, coordinate, and collaborate at work? And can I provide insights that can help them do these things more effectively so that they can reach their collective and individual objectives?* These questions may sound rather simple, and perhaps rather instrumentalist as well. But their answers were changing dramatically due to major changes in how we work: in information technologies, organizational structures, societal expectations, and even in how we commonly define and conceptualize "work" and "workplaces" (Spinuzzi, 2023b). And these dramatic changes were causing obvious and visible stress for people, who spent roughly half of their waking lives at work and who were becoming bogged down in learning unstandardized interfaces and unfamiliar software logics. As a working technical writer who generated software documentation for end users, I saw their distress, and I saw an opportunity to reduce that stress by systematically studying how people worked and carrying lessons back to other technical writers.

Workplace studies were already being conducted in the field, but often with relatively thin theoretical and methodological grounding (e.g., Odell & Goswami, 1985). So, to undertake my work, I sought a conceptual and methodological framework that could guide my studies of workplaces. In 1994, the year I began my Ph.D., my field was mainly borrowing such frameworks from a grab bag of other fields and disciplines: literary studies, cognitive psychology, sociology, anthropology, education, human-computer interaction. In writing studies, these influences (as we'll discuss further in chapter 7) had coalesced into two camps: the social and the cognitive. But some of us were interested in drawing from both; we wanted to tap into a tradition that was *sociocognitive*. And one of my professors, David R. Russell, had become aware of such a tradition. He called it "activity theory."

David soon became my dissertation supervisor, and both of our writings soon became illustrated with various triangles depicting how individuals and small groups used mediators to accomplish their objectives (e.g., Russell, 1995, 1997a). These triangles initially resembled Vygotsky's mediational triangle, with a subject addressing an object (or, as Russell styled it, an "object(ive)") via mediational means. However, I found more elaborated triangles in one of Russell's sources: a second-generation photocopy of Yrjö Engeström's (1990) *Learning, Working, and Imagining*, which he had received from his frequent collaborator Charles Bazerman. In that book were studies of work that resembled the sorts of case studies I wanted to conduct: small sets of people working and talking together, attempting to solve shared problems with tools and texts. The scale, approach, and conceptual tools involved in these descriptive qualitative case studies seemed ideal for underpinning sociocognitive studies of professional communication. They appealed to others in our field as well: Russell (1997b) overviewed such studies emerging in the mid-1990s. In the vein of these studies, I soon began conducting my own qualitative case studies of technologically mediated work, using cultural-historical activity theory (CHAT) to describe and analyze the work of software developers (Spinuzzi, 2003a) and traffic safety workers (Spinuzzi, 2003b). Others in our field did as well.

My case studies drew heavily from Engeström, but also from the burgeoning CHAT studies in another interdisciplinary field that was comparably new:

human-computer interaction (HCI). Studies in HCI, and in affiliate subfields such as computer-supported cooperative work (CSCW), were also frequently small-scale descriptive case studies of work that drew from a sociocognitive perspective, but rather than focusing on the conversation analysis that characterized Engeström's early workplace studies, they also examined and characterized observations of small-scale interactions. Susanne Bødker's 1991 book *Through the Interface* (1991) was a big influence on me, but an even bigger one was Bonnie Nardi's 1996 collection *Context and Consciousness*, in which multiple HCI scholars presented case studies of work (many of which were again illustrated with triangles). Since HCI had several things in common with professional communication—including a focus on how people communicate, coordinate, and collaborate at work, at a time when technology was initiating rapid changes in workplaces—I took plenty of cues from these studies, using them to inform my own study designs, drawing on their insights, and citing them in my own work. In retrospect, these HCI studies became crucial for case studies, since Engeström's own studies became less descriptive, less interested in the minute details of work, and more interventionist as time went on. (Looking back, I realize that the descriptive workplace case studies in *Learning, Working, and Imagining* were all coauthored with Ritva Engeström, whose interest in conversation analysis is evident in another article that deeply influenced me: her 1995 "Voice as Communicative Action").

Nevertheless, Yrjö Engeström's version of CHAT became highly influential in writing studies, especially for guiding and analyzing descriptive qualitative case studies of work. Over the next decade, we saw more triangles throughout our journals, depicting classrooms, workgroups, and organizations, and sometimes linked triangles that represented organizational structures. Terms such as "activities," "activity systems," "activity networks," and "mediators" became common enough that today they are sometimes used without any citations. Engeström's CHAT had become part of the fabric of my field, so much so, in fact, that activity theory, CHAT, and Engeström's CHAT have often been conflated. Literacy researcher Peter Smagorinsky, for instance, complains that he had to stop calling himself an "activity theorist" because reviewers of his manuscripts expected him to be aligned with Engeström (Smagorinsky 2011, p. xxxviii).

In retrospect, when these triangles gusted into our field and settled into our journal pages, they brought along some tribulations as well.

Theoretical and methodological frameworks work by reducing complexity to a set of things we can reasonably manage, and that means selecting what will be important and screening out what won't be. For Engeström's CHAT, what is important—what is named, conceptualized, described, and illustrated with those triangles—is the history and development of labor at individual and collective levels. Consequently, Engeström's CHAT has sometimes been criticized as providing a managerial view: one that depicts how an organization's many resources (people, tools, rules) are arranged to reach a managerial objective. To foreground this focus on cyclical collective labor, Engeström's CHAT backgrounds many other considerations, such as power, persuasion, and affect. Such considerations are key to *rhetoric*, a critical strand in writing studies. Rhetoric, which we can gloss as the study of persuasion and argumentation, focuses on how and why people accept, take up, and act on propositions. As a professor of rhetoric and writing, I am keenly aware of CHAT's limitations for addressing rhetoric, including the tendency for CHAT studies of workplace writing to take a managerial viewpoint.

A contrasting, rhetorical view is provided by the sociology of translation. Michel Callon (1986a) astutely argues that, in being taken up, propositions are also translated, altered so that they can address the interests of an associative network. For instance, a technical innovation does not "sell itself"; it is not an objectively good idea for everyone. Rather, an innovation spreads because those who take it up find ways for it to be useful for their purposes. And in being taken up, the innovation is enacted differently, changed to meet new needs and strengthen new relations. It becomes persuasive by being altered to fit the new network of associations.

We see this dynamic in studies of science (e.g., Latour 1988), technology (e.g., Akrich, 1993; Callon, 1986b; de Laet & Mol, 2000; Latour, 1996), and entrepreneurship (Akrich et al., 2002a, 2002b; Spinuzzi et al., 2016, 2018). But—I argue—we also see it in how Engeström's CHAT was taken up in writing studies, as discussed in chapter 7. In terms of theory, we amplified and developed linkages between CHAT and genre theory, which provided a direct treatment of complex texts that CHAT did not. In terms of methodology, we mainly focused on small-scale descriptive workplace studies rather than implementing the interventionist change laboratory methodology that Engeström emphasized throughout most of his career. And in terms of development, although we periodically imported theoretical developments from Engeström's CHAT, we sometimes interpreted these differently—and we added our own embellishments. The result was, perhaps, heterodox. But that is to be expected: *to translate is to betray*, as Bruno Latour tells us (1999, p. 87). Taking up CHAT meant adapting it to our field's needs, changing it into something that worked for us.

Critically, we in writing studies weren't the only ones to translate/betray Engeström's CHAT. In fact, as I began to read more deeply into the history and development of Engeström's CHAT, I realized that it could be seen as a series of such translations—not as a linear succession of increasingly complex generations, as Engeström often describes it (e.g., Engeström & Sannino, 2021), but as diverse translations by different interested parties who came in and out of a larger ongoing conversation. In understanding the history and development of CHAT *rhetorically*, as a series of propositions that were taken up at different points in different associative networks with different levels of success, we can better understand how this framework was honed for specific purposes—specific arguments. We can step back and see Engeström's CHAT emerging from a set of translations, just as we might analyze the translations that yielded more prosaic innovations: vaccines (Latour 1988), transportation systems (Callon, 1986b; Latour, 1996), bush pumps (de Laet & Mol, 2000), and gazogenes (Akrich, 1993).

To trace and examine these rhetorical translations, let's take a detour to examine two of those emblematic triangles.

A TALE OF TWO TRIANGLES

Let's consider these two triangles, drawn fifty-seven years apart (figures 1.1 and 1.2).

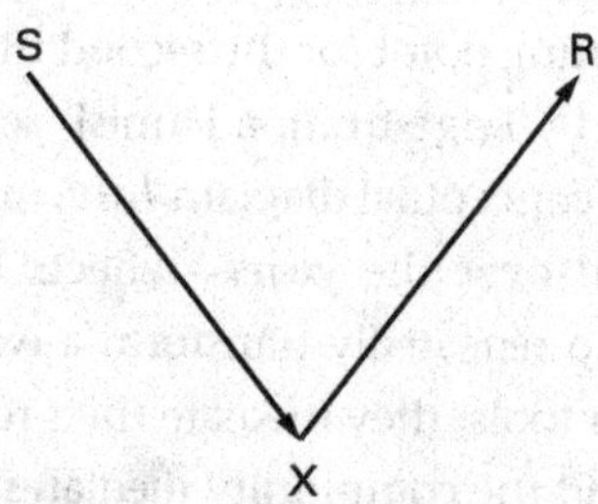

FIGURE 1.1
Mediated action (based on Vygotsky 1978, p. 40)

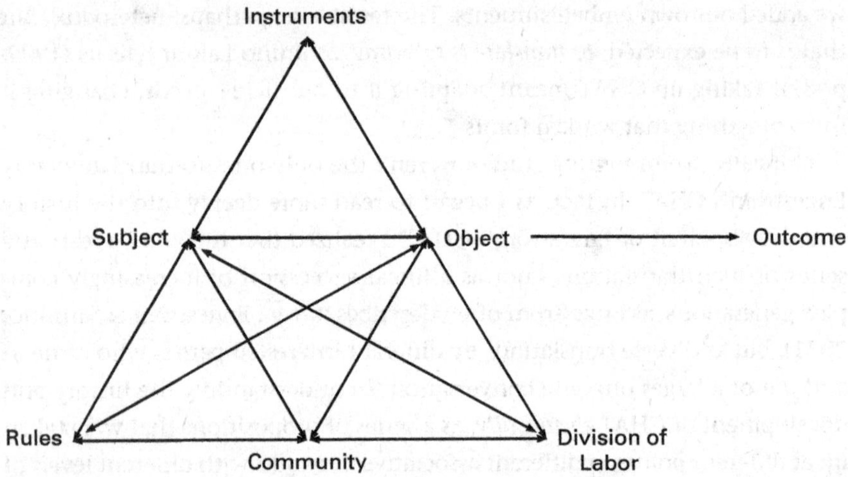

FIGURE 1.2
The activity system (based on Engeström 1987/2014)

The first triangle (figure 1.1) was drawn around 1930 by a young Soviet psychologist named Lev Vygotsky.[1] This conceptual diagram reflects Vygotsky's contention that in human beings, the relationship between stimulus and response is mediated by "psychological tools," that is, signs. For instance, if a Russian man of Vygotsky's time wanted to remember a task he had to do later, he might tie a knot in his handkerchief. When he reached for his handkerchief later, he would discover the knot—a second stimulus—and remember to perform the task. The knot functions as a mediator through which the individual controls his own future actions. Vygotsky tied this self-control to labor, referring to Engels' account of how humanity emerged from labor activity (Vygotsky & Luria, 1930/1994, p. 165; cf. Vygotsky, 1930/1994b).

Labor was also the starting point for the second triangle (figure 1.2), which was drawn around 1987 by Engeström, a Finnish scholar who was studying workplace learning. This conceptual diagram—the most famous of many that Engeström has deployed over the years—reflects Engeström's contention that when actors strive to repeatedly transform a work object, they mediate that transformation with tools, they mediate their relationship to their community through rules, and the community mediates its relationship with the work object via a division of labor. Through these three mediators, the work activity persists and the collective learns.

What's the relationship between these two triangles, these two conceptual diagrams, which both describe human learning and development? Well, it's complicated.

Engeström and his colleagues see the second triangle as an *extension* of the first: they understand it as taking Vygotsky's initial insights of mediated action and scaling them to address larger social systems, that is, collective activity. That is, the second triangle is a version of the first one, aligned closely with Vygotsky's insights but developed further, via a series of innovations. It has been productively *translated* from a theory of individual learning, applied instead to the question of learning in and across organizations. In that capacity, it has been used to analyze interactions, identify systemic issues, examine innovations, conduct workshops, redesign work tools, and develop coordinative texts in a range of areas, including health care (Engeström, 2018), law (Engeström, 1992), information systems (Karanasios & Allen, 2018), human-computer interaction (Kaptelinin & Nardi 2012), telecommunications (Spinuzzi, 2008), organizational learning (Guile, 2010), and public policy (Sannino, 2020a, 2020b).

But some critics instead see the second triangle as a *betrayal* of the first. For instance, Ronald Miller complains that modern uptakes of Vygotsky's work have done violence to it: "There is a not-so-thin line between interpretation and misrepresentation and it seems that this line is increasingly being ignored" (Miller, 2011, p. xi). He professes bafflement at the origins of Engeström's triangle, stoutly arguing that it is not the stimulus-response triangle that Vygotsky uses (pp. 214–221) and that it is more of an "article of faith" than an explanatory device (p. 221). He also bitterly charges that "it is probably fair to say that no other diagram in the history of psychology has been reproduced as frequently and uncritically as this one" (p. 220). Another critic, Andy Blunden, characterizes Engeström's triangle as a "handy pocket manual of social analysis" but not a concept per se (2010, p. 232); whereas Vygotsky's triangle represents a fundamental unit of analysis, Engeström's triangle does not (a question to which we will return in chapter 11). Still another critic, David Bakhurst, complains that whereas Vygotsky and others in his tradition "saw the concept of activity as a fundamental category to address profound philosophical questions about the possibility of mind," "activity theory in the West has principally become an empirical method for modeling activity systems" through figure 1.2's triangle diagram

(2009, p. 197). Building on Bakhurst's critique, Alex Levant characterizes Engeström's triangle as "emblematic" (2018, p. 103), and although he allows that Engeström's work is useful, it focuses on "mundane pursuits" rather than the critical, fundamental questions Vygotsky explored. Finally, discussing activity system modeling, Anna Smith and Paul Prior argue that the triangle "seems to envision the social in vertical, institutionally-anchored terms that privilege the institution's perspective (e.g., in the routine identification of the subject of activity systems as the official, front-stage representatives of the institution—doctors in hospital, teachers in schools)" (2020, p. 3). They see Engeström's triangle as producing not a *critical* perspective of labor activity but a *managerial* one.

Engeström's triangle, then, functions as a sort of Bermuda triangle in which supporters and critics have dramatically different encounters, one in which Vygotsky's insights about mediation either disappear or sail through intact. For both camps, it functions not just as a conceptual diagram but also as an emblem that denotes allegiance: not just in this matter but in a cluster of related terms, concepts, empirical focuses and scopes, and methodologies. For instance, Vygotsky and Engeström both use terms such as "mediation," "learning," "zone of proximal development," and "double stimulation," but these are operationalized and arguably conceptualized differently. They are also scaled differently: whereas Vygotsky applied them to individuals, usually children functioning in dyads with more experienced individuals in settings such as schools and rehabilitation facilities, Engeström applied them to work groups, organizations, and publics. The triangle, although roughly the same size on the page, is greatly expanded in spatial, temporal, and social dimensions. Thus, not just in its triangle, but also in its vocabulary and concepts, Engeström's work is often seen in terms of heterodoxy: depending on the observer, it has either *advanced* or *abducted* Vygotsky's thought, applying it to a different scale of social phenomena.

Is Engeström's triangle a progression of Vygotsky's triangle or a deviation? A translation or a betrayal?

TRANSLATING—AND BETRAYING—CHAT

The French sociologist Michel Callon argues that to translate is to betray (1986a, p. 19). Changing something, even something as small as a triangle, can be understood both as a further assertion (since we are continuing to develop that thing) and as a rejection (since we find the previous iteration inadequate). Adapting something helps it to persevere in the face of changes, but it also necessarily changes that thing.

Case studies of translation have been on offer since before Engeström formulated his triangle. Yet these case studies have mainly happened in science and technology studies (STSs), where researchers have examined translations in biological and physical sciences (Latour & Woolgar, 1979; Latour, 1987, 1988), medicine (Mol, 2002; Law & Singleton, 2005), and technical innovations and systems (Latour, 1996; Law, 2002; de Laet & Mol, 2000; Akrich, 1993; Spinuzzi, 2008). That is, translation has mainly been deployed in the "hard" sciences, especially in Western Europe and North America, where we can often see concrete changes and chains of re-representations in scientific data, technological artifacts, and infrastructure.

In contrast, we don't see as many translation studies focusing on the social sciences: psychology, sociology, cultural anthropology, and other studies of human behavior, where the innovations tend to be in techniques, methodologies, and concepts (see chapter 2). But translation happens in these "soft sciences" as well—as our comparison of the two triangles suggests. Like Brahe, Boyle, Pasteur, and Guillemin, Vygotsky and Engeström produced inscriptions to re-represent their object(s) of analysis—in this case, social and psychological relationships, ranging from an *aide-memoire* that fits in one's pocket to a global phenomenon[2] that envelopes the Earth, like climate change—in ways that made these relationships tractable: "They are made *flat*. There is nothing you can dominate as easily as a flat surface of a few square meters" (Latour, 1986, p. 19).

In our triangle case, however, different camps disagree about what, if anything, is being "dominated" in the second triangle. These triangles are in the same family, and there's no fight as vituperative as a family fight.

The "family" is cultural-historical activity theory (CHAT), a term that emerged in the early 1990s (see Ekeblad, 1999) to express a relationship between *cultural-historical psychology*, which takes Vygotsky as its starting point, and *activity theory*, which similarly claims Vygotsky as a starting point but traces its origins through Vygotsky's associate A. N. Leontiev. American psychologist Michael Cole, an early popularizer of CHAT (e.g., Cole, 1996), learned about both Vygotsky's cultural-historical psychology and Leontiev's activity theory from A. R. Luria, who was Vygotsky's close associate as well as a lifelong associate of Leontiev. Like Vygotsky, Leontiev, Luria, and Cole,

CHAT in all its variations is concerned with social human beings learning and developing.

As we'll see in chapter 3, CHAT has many variations, sometimes described using metaphors such as generations, strands, or waves. These variations have some common tenets that make CHAT well suited for examining learning and development. But human beings learn and develop in different ways and within different social arrangements—for instance, primary schooling, rehabilitation, bureaucratic work, public works, and spontaneous collaboration across otherwise unaffiliated specialists. These social arrangements have different power dynamics, expectations, time scales, and scales of interaction. They are arguably too much to fit into a single triangle, no matter how complex or expansive. Hence, we see different variations, often drawing concepts and methods from other social science approaches. Consequently, these variations have persistent disagreements that often revolve around each variant's fidelity to the great figures of its past: Is this variant consistent with Marx? With Engels? Vygotsky? Luria? Leontiev? Ilyenkov?

Our focus in this book is a specific variant that has been taken up by many scholars working in many areas, including (and this is just a short list) psychology, education, sociology, philosophy, workplace learning, management studies, human-computer interaction, cooperative work, and—as we've seen—writing studies. This variant, developed and popularized by Yrjö Engeström, is sometimes called Engeströmian activity theory, Finnish activity theory, or Scandinavian activity theory. For brevity, I will simply call it "Engeström's CHAT," emphasizing Engeström's pivotal role in developing it. Engeström's CHAT emerged from dramatic shifts and reconceptualizations, translations, and, inevitably, betrayals. These translations and betrayals go well beyond triangles, encompassing issues of theoretical and methodological precepts, scope, agency, application, and field.

And they continue as Engeström's CHAT continues to adapt. For example, thirty-four years after figure 1.2 and ninety-two years after figure 1.1, Engeström and Sannino produced another figure, one with multiple intersecting loops representing different institutions (Engeström & Sannino, 2021, p. 15). In that figure, the triangles disappear altogether: Engeström and Sannino have moved from a study of the internal relations of working groups to a much broader scale, the scale of institutions that co-learn as they attempt to implement a public policy. As Latour notes, "No matter what the (reconstructed) size of the phenomena, [inscriptions] all end up being studied only when they reach the same average size" (1986, p. 19). CHAT's diagrams have not gotten bigger on the page, but they portray vastly larger spatial, temporal, and organizational scales. Engeström's CHAT has moved a long way from its roots in Vygotsky's studies of children working briefly in dyads with their parents and instructors.

Critically, Engeström's CHAT was not fated to develop in this way, and the second triangle was not fated to emerge from the first. Rather, we can trace transformations in the CHAT literature, looking at key instances in which CHAT was taken up, elements were translated, and the result was successfully taken up again elsewhere. That is, Engeström's CHAT has been produced not just through systematic analysis and synthesis, but also through a series of political and rhetorical transformations—and, yes, even some haphazard accidents and sleight of hand—as it has been deployed under different political, cultural, and social conditions for different purposes. These transformations, sedimented as a series of settlements, have allowed it to persist as a social science approach and develop as a framework for workplace studies.

Such transformations have been explained elsewhere using different metaphors: as a linear set of generational transformations or as a bundle of internally coherent strands of thought that are applied to different problems and literatures or as subsequent waves of rereadings (see chapter 3). These are all developmental metaphors that understand CHAT in terms of CHAT. In this book, however, we will understand it in terms of translations. And, critically, we will not argue that the translations leading to Engeström's CHAT were necessary or fated or ideal; we will instead examine how these translations responded to conditions by positioning CHAT differently at specific points, adapting it so it could be taken up further elsewhere.

A full history of translations would be impossible, but we will examine specific points of translation that have made Engeström's CHAT what it is today. These translations are sometimes shown in figures such as triangles (see figure 1.3 at the end of this chapter), but they are often shown in other ways: how terms are picked up and used differently, how cases are selected, how literature is drawn upon, and what phenomena are examined and explained. As we examine these translations, we will answer these questions:

- How was Engeström's CHAT accomplished? That is, how did this variant of cultural-historical activity theory emerge through successively addressing different problems? More prosaically, how did we get from the 1930

triangle to the 1987 triangle and beyond? How did CHAT generate, use, and progressively alter the figures compiled in figure 1.3?
- What does this story tell us about how social science frameworks spread?
- What lines of critique, including self-critique, can this analysis open for us?

That third question is important for me personally. I've been using Engeström's CHAT for over twenty-five years to guide and analyze my studies of workplace communication, and I fully plan to continue. So this book is not an exposé or an unmasking (although it is arguably a gossipy tell-all). At the same time, it's certainly not a triumphalist story of how one figure imposes their will on others. Quite the opposite. It's a book about how innovations spread by being taken up and repurposed, by developing theoretically and analytically.

A WORD ON THEORY AND ANALYSIS

Before going any further, let's establish the terms "theory" and "analysis" as I use them in this book.

"Theory" is often understood in the natural sciences as a scientifically established principle that explains phenomena, especially one supporting a predictive model. But in this book, I'll instead use it the way it is often used in the social sciences: there, a theory is "a big idea that organizes many other ideas with a high degree of explanatory power" (Collins & Stockton, 2018, p. 1), while a theoretical framework is "the use of a theory (or theories) in a study that simultaneously conveys the deepest values of the researcher(s) and provides a clearly articulated signpost or lens for how the study will process new knowledge" (p. 1). Similarly, Miles et al. say that "a theoretical framework utilizes theory/theories and their constituent elements as the presumed 'working model' that drives the investigation and analysis of a social phenomenon" (2020, p. 15). This is how the term "theory" in the phrase "activity theory" has been understood: "Activity theory is a clarifying, orienting framework. It is not a 'theory' in the traditional sense in which theory is understood in natural sciences. Activity theory does not support creating and running predictive models which only need be 'fed' with appropriate data. Instead, it aims to help researchers and practitioners orientate themselves in complex real-life problems, identify key issues that need to be dealt with, and direct the search for relevant evidence and suitable solutions" (Kaptelinin & Nardi, 2012, pp. 6–7).

Similarly, "analysis" is often understood in the natural sciences as examining some phenomenon to identify essential, enduring components. However, in qualitative research, analysis is typically understood as interpretive, with categories emerging inductively from the ongoing analysis (Basit, 2003; Corbin & Strauss, 1990). Bernard says that "analysis is the search for patterns in data and for ideas that help explain why those patterns are there in the first place. . . . It starts before you collect data—you have to have some ideas about what you're going to study—and it continues throughout any research effort. As you develop ideas, you test them against your observations: Your observations may then modify your ideas, which then need to be tested again, and so on. Don't look for closure in the process. If you're doing it right, it never stops" (Bernard, 2011, p. 338). Analytic categories are thus understood as contingent and situationally useful.

With that grounding, let's get to the rest of the book.

THE REST OF THIS BOOK

In the rest of the book, we'll undertake these questions in three parts.

In part I, "Method and Research Object," we'll establish the fundamentals of our analysis: the analytical method we'll use and the research object we'll explore.

In chapter 2, "Translations and Betrayals," we will establish our *analytical method,* the sociology of translation. We'll review its foundational concepts, its application, and its value for understanding our research object and answering our research questions. Although the sociology of translation has been used extensively to understand the physical sciences, it has been applied more rarely to the social sciences; here, we'll establish its strengths for taking on our task.

In chapter 3, "CHAT's Generations—or Strands, or Waves?" we will establish our *research object*: Engeström's CHAT. We'll review its basic concepts and how it is being applied. We'll do this in part by examining three metaphors that are commonly used to describe its development: generations, waves, and strands. CHAT is a developmental theory, so we should not be surprised that its practitioners describe its history as developmental—but we will treat these stories of development skeptically, and we will set the stage for applying translation, a nondevelopmental account.

In part II, "Translations of CHAT," we'll examine key points of translation in the history of Engeström's CHAT. A full accounting of that history is impossible—certainly in a book of this size. But these key points allow us to examine specific and consequential moments of translation, moments that became sedimented into Engeström's CHAT as we know it today. In each chapter, we'll discuss the history of CHAT during a specific period and then closely examine how that era of CHAT was translated in three key categories: *empirical focus*, *mastery*, and *mediators*.

Chapter 4, "Vygotsky's CHAT: Marxist Psychology before the Great Terror," examines the origins of cultural-historical psychology in the works of Vygotsky and his circle in the early Soviet Union, recounting the trials they faced as they struggled to develop a genuinely Marxist psychology. This story includes their formulation of the foundational concept of mediation by signs (or "psychological tools") as a way to bring about the envisioned New Soviet Human as well as their application of this theory in education, rehabilitation, and a stymied attempt at cross-cultural psychology. Vygotsky's works and phases have been discussed extensively in the literature, but by examining this period in terms of translation, we can identify how Vygotsky's framing interested different actors in his cultural-historical psychology. Yet the story ends in tragedy, with Vygotsky dead and his disciples denying him under the shadow of the Great Terror of 1936–1938.

Chapter 5, "Leontiev's CHAT: Marxist Psychology after the Great Terror," picks up the story of Vygotsky's collaborator Leontiev, who broke with Vygotsky's focus on cultural meaning, instead focusing on *activity*, a category imported into psychology from philosophy by Rubinshtein. Leontiev's activity theory account collapsed psychological and physical tools into one category, dodging the accusations of idealism that had dogged Vygotsky, and assiduously grounded the account in the works of Marx, Engels, Lenin, and Stalin. These and other translations allowed Leontiev's activity theory to fulfill the key criteria of science in the Stalinist period: *partiinost'* (party spirit), Marxism, and practicality. Rather than focusing on unlocking the potential of the New Soviet Human, activity theory focused on developing Soviet citizens to serve under the supervision of command socialism. With these alterations, activity theory weathered the trials of the Great Purge and was taken up as a successful psychological approach in Stalin's USSR. Consequently, Vygotsky's collaborators gained greater leeway to revive his legacy—selectively, in ways that affirmed their translations. This story ends in triumph: Leontiev ended his days as dean of the Faculty of Psychology at Moscow State University, with activity theory established as a leading school of psychology in the USSR. But it also ends in a betrayal coming to light.

In chapter 6, "Westernized CHAT in 1987: An Interventionist Design Approach," we'll move the scene to Western Europe in 1987, where two nonpsychologists—Susanne Bødker, a Danish Ph.D. student studying human-computer interaction, and Yrjö Engeström, a Finnish Ph.D. student studying workplace learning—independently picked up CHAT as an approach that could address the deficiencies of the paradigm that was dominant in Western studies in those fields: information-processing cognitive psychology (IPCP). Thus, rather than applying CHAT to psychological studies of individuals and dyads, both Bødker and Engeström used it to underpin interventionist design projects involving small work groups beyond the dyad. And rather than focusing on the development of individuals, CHAT in this iteration shifted its focus to the development of the artifacts and divisions of labor that would mediate these individuals' work. Through these new alterations, CHAT survived the trials of two trips: the trip to the West and the trip from psychology to new fields impacted by the learning sciences. It became poised to analyze, model, and intervene in small-scale work organizations. The price, some argue, was that it gave up its birthright of economic and cultural critique so that it could focus on the more mundane question of incremental changes in work conditions.

Before exploring the development of Engeström's CHAT further, we'll take a short detour in chapter 7, "Interlude: CHAT as a Sociocognitive Base for North American Writing Studies." In this chapter, we'll leave the mainstream of Engeström's CHAT to look at a branching stream, the localized uptake of CHAT in my own field of writing studies. A few years after the events of chapter 6, writing studies was struggling to reconcile cognitive and social approaches to understanding writing in empirical cases. Like many other fields (e.g., human-computer interaction, computer-supported cooperative work, user experience design, information systems), we turned to CHAT, translating it further to meet our own localized needs. Here, I'll describe this uptake of CHAT as a sociocultural framework that could underpin descriptive case studies of writing in concert with genre theory. In taking up Engeström's CHAT, we reoriented it to address scenes in which people produced, interpreted, and interrelated types of texts in the course of their activities. Like many localized translations in other fields, these translations generally did not strongly influence the mainstream of Engeström's CHAT.

We'll return to that mainstream in chapter 8, "Engeström's 'Third-Generation' CHAT: An Organizational Sociology." This chapter picks up a decade after chapter 6, as Engeström's CHAT expands to address larger workgroups, whole organizations, and interorganizational work. Engeström's CHAT, whose roots were in a USSR that had *banned* sociology (Osipov, 2009; Weinberg, 1974), had by this point *morphed into* an organizational sociology. It was applied to a broader variety of organizations in far more fields. This expanded scope also introduced new trials: other social theories and methodologies had already occupied this space and produced competing solutions to its problems. In response to these trials, Engeström and collaborators methodically expanded concepts by borrowing from (and translating) a wide range of theoretical and methodological sources, sometimes by *appropriating* concepts into Engeström's CHAT (such as co-configuration and networks), sometimes by *substituting* concepts more compatible with Engeström's CHAT (such as knotworking, mycorrhizae, and runaway worlds). This period involved a rapid expansion in the conceptual vocabulary of Engeström's CHAT—arguably introducing incoherencies. It also involved a shift in focus in the mediators being studied: rather than examining broad sets of mediators at a granular level, these studies refocus on generating new coordinative genres that could improve the interoperation of intermeshed activities. Thus, these studies crept ever farther away from Marxist studies of dyads and ever closer to an apolitical, managerial perspective.

Chapter 9, "Engeström's 'Fourth-Generation' CHAT: An Implementation of Public Policy," picks up in the wake of Engeström's (2009a) acknowledgement that in social production and peer production, "the boundaries and structures of activity systems seem to fade away" (p. 309). In later publications as well as interviews with other scholars, Engeström and Sannino (Engeström & Sannino, 2021; Ploettner & Tresseras, 2016; Yamazumi, 2020)—stung by Marxist critiques of Engeström's CHAT—argued that the next iteration of Engeström's CHAT should be about *alternatives to capitalism* (cf. Engeström & Sannino, 2016, p. 403). Thus, the efforts described in this chapter addressed learning in social movements, consequently expanding empirically far beyond the scope of Vygotsky's and Leontiev's dyads, the small bounded groups of Engeström's early work, or even the organization-specific change laboratories of Engeström's 1990s work. The addressed social issues included childbirth support, sustainable food cooperatives, social justice, and homelessness—public policy issues, which Engeström and Sannino positioned Engeström's CHAT to solve. And to scale Engeström's CHAT for this new breadth of scope, Engeström and Sannino built on the concept of runaway objects, introduced the concept of heterogeneous coalitions, and repurposed (and transformed)

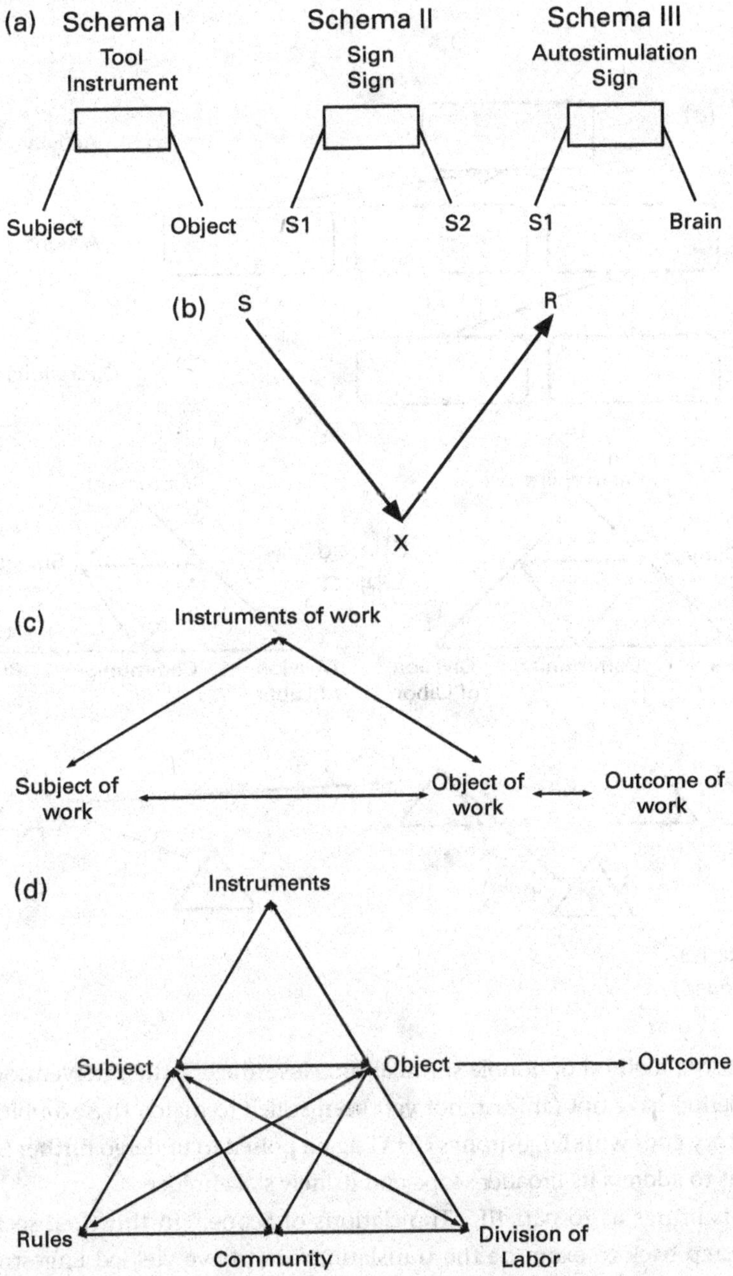

FIGURE 1.3

Triangles used across the history of CHAT. (a) 1929: Tool and sign mediation (based on Vygotsky, 1929/1989, p. 62). (b) 1930: Mediated action (based on Vygotsky, 1978, p. 40). (c) 1984: Mediated work action (based on Engeström & Sannino, 2021, p. 7. (d) 1987: Activity system (based on Engeström, 1987/2014). (e) 1987: Levels of individual human activity (Bødker, 1987, p. 28). (f) 1990: A network of two activity systems (based on Engeström, 1990, p. 128). (g) 1992: A network of many activity systems (loosely based on Engeström, 1992, p. 13).

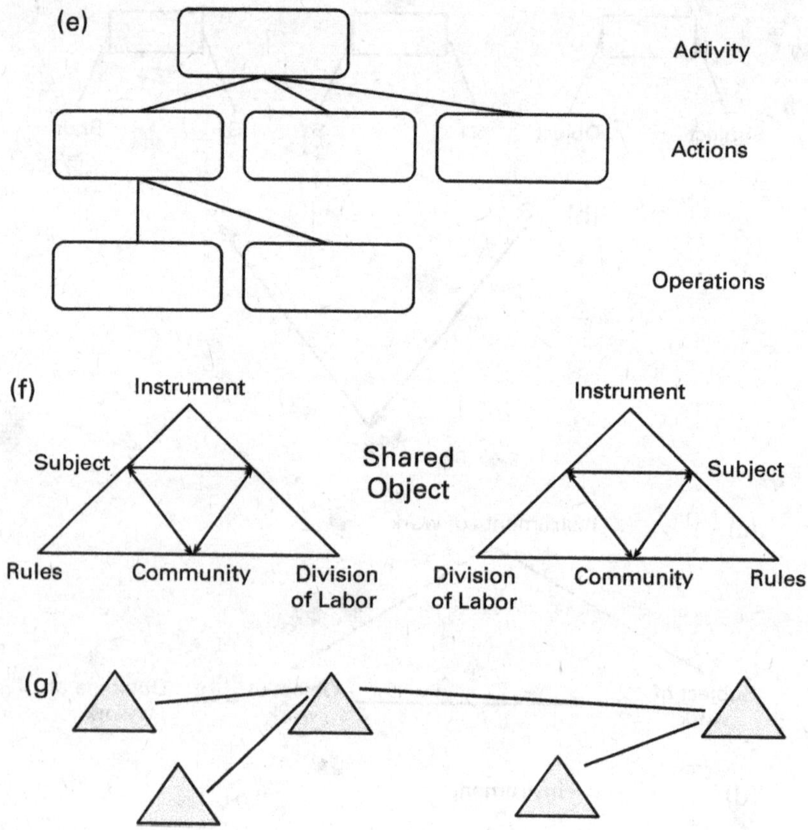

FIGURE 1.3
(continued)

Vygotsky's method of double stimulation. Nevertheless, the interventions in this period have not (at least, not yet) been scaled to match these ambitions. The story ends with Engeström's CHAT again poised to undergo further translations to address its broader scope and its new stakeholders.

This brings us to part III, "Translations of Scope." In this final section, we'll step back to examine the translations that have yielded Engeström's CHAT, drawing broader lessons about what a translation approach can tell us about the social sciences.

In chapter 10, "The Scope of the Expanding Object," we'll look at the developmental trajectory traced by the translations in empirical scope that we saw in part II. As part of its alteration, Engeström's CHAT has expanded rapidly and remarkably, attempting to cover ever more concepts and context as it has been applied to ever-widening cases. Put crudely, Engeström's CHAT has been deployed as a theory of context, yet the nature of context is that there is always more of it, an ever-receding horizon for Engeström's CHAT to reach. It has chased these larger contexts by identifying larger objects, and it has investigated these objects by expanding both theoretically and methodologically.

Chapter 11, "The Scope of the Unit of Analysis," addresses the oft-mentioned but seldom-discussed question of the unit of analysis. Vygotksy's and Engeström's triangles are often described as representing units of analysis for sociocultural research. But what does that actually mean? In this chapter, we'll review what a unit of analysis is and what additional requirements people have tried to apply to a CHAT unit of analysis. We critique the various units that have been applied throughout the history of CHAT and examine how they track CHAT's translations.

Finally, in chapter 12, "The Diffusion and Translation of CHAT," we'll discuss the implications of this translation analysis, both for Engeström's CHAT and for translation studies in the social sciences.

I METHOD AND RESEARCH OBJECT

2 TRANSLATIONS AND BETRAYALS

Let's consider the fable of the scorpion and the frog, a fable that (like CHAT) seems to have emerged in Russia shortly after the Revolution. In this fable, a scorpion asks a frog to transport it across the river on the frog's back. The frog initially refuses, worrying that the scorpion will sting it. But the scorpion promises not to: after all, if it does, they will *both* drown! Reassured, the frog agrees. Yet, in the middle of the river, the scorpion does sting the frog, condemning both of them to death. *Betrayal was in its nature.*

As I hinted in chapter 1, CHAT's history is full of betrayals and accusations of betrayal. Here are just a few examples:

> In 1923, just a few years after the Bolshevik Revolution, academicians and scientists are attempting to reconceive all subjects in Marxist terms. Georgy Chelpanov, who organized a graduate program in psychology at Moscow University, is condemned and deposed as director of the State Institution of Experimental Psychology, or GIEP, by his graduate student Konstantin Kornilov (McLeish, 1975, p. 108; Joravsky, 1989, p. 224). Kornilov proffers his own approach, called reactology, as a truly Marxist psychology. But reactology needs allies, especially against established psychologists such as Pavlov and Bechterev. So when a young psychology teacher named Vygotsky speaks at the Second Psychoneurological Congress in 1924, arguing eloquently that Marxist psychologists could study consciousness without being idealists, Luria (then a student) persuades Kornilov to hire him. Kornilov does, and directs Vygotsky's dissertation, which uses reactology to analyze responses to literature. Kornilov even waives the dissertation defense for Vygotsky, who by 1926 is bedridden in a hospital after a bout of tuberculosis. Despite this kindness, at exactly this point, Vygotsky abandons reactology. He spends his hospital stay in 1926–1927 writing a book that sketches a new psychology—not reactology, but what he regards as a fully Marxist psychology in which signs allow people to mediate their

own actions. By 1930, reactology has fallen out of favor, and Kornilov himself is deposed as director in 1931.

From Kornilov's perspective, Vygotsky was recruited to support reactology, but even after Kornilov showed him great kindness, Vygotsky defected and undermined it. This was a professional and personal betrayal. But then again,

> in 1930, Vygotsky produces several works on his instrumental psychology. But by 1932, Vygotsky concludes that the instrumental approach is not adequate, and in a stormy meeting with his circle on December 4–5, he criticizes his earlier instrumentalist work, arguing instead for "the principle of the semantic structure of the mind": he retains sign mediation but "considerably broaden[s] its concept by transitioning from the study of the external structure of the sign operation in its instrumental function to the study of its internal structure, i.e., the meanings" (Zavershneva & Van der Veer, 2018, p. xvii). Vygotsky's associate Leontiev strongly disagrees: he and his Kharkov Circle want to keep building on the foundation of instrumental psychology that Vygotsky has laid. Furthermore, they think Vygotsky's focus on signs and language is misplaced. In private correspondence, Leontiev characterizes Vygotsky as reducing all psychology to word meaning (Zavershneva, 2016a) and instead argues for replacing the emphasis on signs with labor activity (Van der Veer & Valsiner, 1991, pp. 289–90), as famed Soviet psychologist S. L. Rubinshtein has done (Payne, 1968). In response, Vygotsky characterizes Leontiev's work as a step backward. This split is still ongoing when Vygotsky dies of tuberculosis in 1934. In one of his last notebook entries, Vygotsky compares himself to Moses: "This is the final thing I have done in psychology—and I will like Moses die at the summit, having glimpsed the promised land but without setting foot on it." (Zavershneva, 2016b, p. 126)

From Vygotsky's perspective, his allegiance was to a truly Marxist psychology—that is the "promised land" to which he intended, like Moses, to lead his people. When reactology failed the test, he developed the instrumental approach. But he judged that instrumental psychology, though Marxist, did not adequately treat psychological phenomena. He tried a new, more fruitful approach—but Leontiev and the Kharkov Circle defected, refusing to follow him. Again, this betrayal was not just professional but personal. But then again,

> at a 1931 conference, A. A. Talenkin condemns Vygotsky and Luria's concept of "psychological tools" as non-Marxist, arguing that the development of mental processes must be grounded in labor (Keiler, 2018, p. 101). Such criticisms cannot be brushed off: Leontiev's first book is delayed for two years and only released with a required self-criticism enclosing it (pp. 95–96). Leontiev pleads with Vygotsky

to blunt these criticisms by revising his theory, but Vygotsky does not listen. In July 1936, two years after Vygotsky's death, the other shoe drops: the Central Committee condemns pedology—the interdisciplinary study of the child, with which Vygotsky had been associated—and calls for severe criticism of pedologists and their books. This call is answered throughout the rest of 1936 and into 1937, with authors condemning not only the dead Vygotsky but also his living associates, including the suddenly vulnerable Leontiev and Luria. The two scramble to protect themselves by criticizing Vygotsky for idealism in their publications and speeches, by distinguishing their own approaches from Vygotsky's, and eventually by associating their research agendas with whichever scientist is in favor at the time: Rubinshtein, Pavlov, and even (humiliatingly) the pseudoscientist Trofim Lysenko.

From Leontiev's perspective, the instrumental approach was not just a sound approach; it was a safe one. It fit easily into Soviet orthodoxy, and it allowed the Kharkov Circle to avoid even the whiff of idealism. By collapsing tool and sign, and by describing psychology as emerging from social relations surrounding labor, Leontiev could ground his work in the unassailable work of Frederich Engels—protecting himself and the rest of his circle. Yet here at the beginning of the Great Terror, Vygotsky wanted to betray this safe approach with one that was clearly vulnerable to charges of idealism. No wonder Leontiev was upset: even after Vygotsky's death, their association nearly discredited Leontiev, who had to work hard to distance himself from Vygotsky's dangerous ideas and haphazard borrowing from bourgeois sources. (Later, Leontiev even praised the Pedology Decree that almost sank his career.) Leontiev and Luria both became adept at sensing the shifting currents and aligning themselves with ascendant scientists while preserving their own approaches. It was only after Stalin's death that they judged it safe to cite Vygotsky again—and Leontiev still portrayed Vygotsky as only a forerunner with some dangerously idealist views.

Unlike the scorpion in the fable, the people in these stories did not betray each other because it was in their nature. Yes, when someone deviated from their path, they saw it as a betrayal. But when they followed their own path, deviating from someone else, they were being faithful to their cause. What seemed to Person A as a betrayal seemed to Person B as a *translation*: a persuasive innovation, a slight detour, without which their path could not continue. Total fidelity would have resulted in failure—existential failure. Vygotsky would not have developed (really, codeveloped) the many insights about mediation, learning, and meaning that are attributed to him. Leontiev would not have climbed to a powerful, influential position in Soviet psychology, which he eventually used to rehabilitate a version of Vygotsky. Engeström and others in the West would not have been able to challenge dominant Western understandings of information-processing cognitive psychology. And so on. These translations, these betrayals (they are, after all, the same thing) allowed a relatively coherent thing called CHAT to emerge and persevere.

I could tell a dozen other such stories of betrayal, family fights, misunderstandings, deceptions, and drama—and I will in part II. But to do so, I can't use a CHAT framework. CHAT is not good at analyzing itself in these terms: it uses developmental metaphors that lack the vocabulary for understanding the political and rhetorical dimensions of its own history (see Spinuzzi 2008). (We'll see and critique three of these developmental metaphors in chapter 3.) Instead, in this chapter, I turn to an analytical framework that is all about such betrayals: the sociology of translation.

THE SOCIOLOGY OF TRANSLATION

The sociology of translation, also known as actor-network theory (ANT), has been used extensively in science and technology studies to examine how science becomes settled. Like CHAT, it has a vast vocabulary attached to a vast set of concepts. I'll draw heavily on the early works of ANT here. In fact, the model for part II is Latour's *The Pasteurization of France* (1988), in which Latour strictly examines three scientific journals for their contemporaneous characterizations of Louis Pasteur's work. In that book, Latour does not take a Great Figure view of this work: he emphasizes that we should not confuse Pasteur with what is attributed to him (p. 15). I'll follow this dictum as well. Even though I'll ground the history chapters in the stories and work of specific principals (Vygotsky, Leontiev, Bødker, Engeström, Sannino), I'll treat these specific instances as samples of larger movements and I'll note when a forerunner's role is retroactively enlarged by the later attributions of other actors. That is, I'll focus on these figures not because they were great, but because a great deal has been attributed to them. (Please accept my apologies for not mentioning or barely mentioning a host of others who were involved in these movements, including Bergson, McCarthy, Lewin, Rubinshtein, Bozhovich, Gal'perin, Davydov, Zaporozhets, Gonzales Rey, and the list goes on and on.)

In these larger movements, we see points at which material objects are transformed. Unlike CHAT, translation focuses on the process of negotiating and transforming such objects to persuade, to gain agreement across human and nonhuman actors; if they are successfully translated, these objects can anchor a rhetorical settlement across these actors. Thus, translation focuses on relational interactions that, unlike dialectical ones, can be reversed. Such settlements are contingent and stabilized for now, representing composite goals (Latour, 1999) and compromises that can fall apart. In these settlements, *a translation is always a misunderstanding*, since the parties to the settlement always have different long-term goals (Latour, 1988, p. 65). Kornilov and Vygotsky (and many others) settled on the goal of a Marxist psychology but had different criteria that eventually led them to split. Vygotsky and Leontiev (and their broader circles) pursued that Marxist psychology through an instrumental approach but came to different conclusions about how well that solution worked. Settlements come together, then often fall apart.

SETTLEMENTS

To understand how such settlements coalesce and collapse, let's turn to the famous case of St. Brieuc Bay that Michel Callon (1986a) describes. In that case, marine researchers, fishermen, and scallops formed an uneasy, mutually beneficial, materially instantiated alliance. The researchers wanted to study scallops to determine how to best preserve them in the long term; the fishermen wanted to ensure their own long-term livelihood by avoiding overfishing; and the scallops presumably wanted to live undisturbed. The researchers coordinated an alliance in which they deployed larva collectors into which the scallops could settle to be studied in safety while the fishermen fished elsewhere. That alliance held until Christmas Eve, when it suddenly collapsed: because scallops fetched high prices on Christmas Eve as people demanded them for the Reveillon feast, the fishermen decided to follow their short-term, individual interests and betray the alliance by raiding the larva collectors. In his case study, Callon examines the alliance in terms of translation: "a definition of roles, a distribution of roles and the delineation of a scenario" (Callon, 1986a, p. 26). By applying the principle of translation, Callon could analyze the nonhuman scallops as participating in this settlement as well. Specifically, he examined four movements:

- **Problematization**: In which a problem or activity is described "in terms specific to a particular discourse" (Swarts, 2011, p. 278). The marine researchers ask, How do we preserve scallops' long-term survival?

- **Interessement**: In which one actor's interests are related to another's. *Intéresser* means "to interest" or "to concern." Like Callon, I use "interessement" throughout this book as a synonym for interposition. In interessement, an actor positions itself in between the other actor and its goals. The marine scientists offer both the fishermen and the scallops a way out of overfishing.
- **Enrollment**: In which actors with related interests are coordinated—a proposed alliance. The marine researchers propose larva collectors and no-fishing zones.
- **Mobilization**: In which a "spokesperson" represents the "consolidation of interests of the group" (Swarts, 2011, p. 278). The collective forms a firm agreement.

Latour similarly argues that "interests are 'translated.' That is, when their goals are frustrated, actors take detours through the goals of others, resulting in a general drift, the language of one actor being substituted for the language of another" (1999, p. 87). The result is a relatively durable settlement in which different actors' interests are knotted together—for a time. Such settlements can always fall apart as the different actors' interests diverge. Callon's case provides one vivid example, but we already saw others at the beginning of the chapter. For instance, let's take Kornilov:

- **Problematization**: Kornilov and many others propose a "Marxist psychology," an absolutely necessary development for psychology to continue as a science in Bolshevik Russia.
- **Interessement**: Luria, who is aligned with Kornilov, hears Vygotsky speak and realizes that Vygotsky can contribute to such a psychology. Kornilov sees his path to a Marxist psychology as (in small part) running through Vygotsky.
- **Enrollment**: Kornilov proposes to hire Vygotsky; Vygotsky accepts—seeing his own pathway to a Marxist psychology as running through Kornilov.
- **Mobilization**: Vygotsky writes his dissertation under Kornilov, using his vocabulary and concepts, producing a work in the vein of reactology. Through these efforts, reactology becomes more real, more elaborated, and a better candidate for "the" Marxist psychology.

Yet that settlement soon fell apart, not because Vygotsky gave up on the mutual problem, but because Vygotsky no longer saw the path to its solution as running through Kornilov. Rather than being a necessary detour between Vygotsky and his goal, Kornilov was now considered to be an *unnecessary* detour, and he soon became disintermediated.

Although I have initially focused on the human actors in these cases, nonhuman ones are active as well. These include tools, rules, signs, labor, dialectic, pedology, the New Soviet Human, higher mental functions, and "humans" who have been re-represented as symbols—including Marx, Engels, Lenin, Stalin, Lysenko, and Pavlov (and, obviously, Vygotsky after his death). Each story told in part II involves enrolling and mobilizing such allies to solidify a given settlement. Settlements are built on top of other settlements: "Constructing a phenomenon in successive layers renders it more and more real within a network traced by the displacements (in both senses) of researchers, samples, graphics, specimens, maps, reports, and funding requests" (Latour, 1999, p. 76). Facts are real *because* they are constructed (p. 275).

For that reason, "For the world to become knowable, it must become a laboratory" (Latour, 1999, p. 45). In the lab, "the weak becomes strong simply by changing the point of application of their efforts" (Latour, 1988, p. 47). For Vygotsky, the move was from the unobservable psyche to the observable actions of children in dyads; for Engeström, the move was from the vast and largely unobservable workings of institutions to observable, key encounters and the participants' dialogue about them. A laboratory is like a map: it relies on "the indefinite sedimentation of other disciplines, instruments, languages, and practices" (Latour, 1988, p. 30), but once it is produced, it can represent just the parts of reality that are relevant for constructing a given fact. Science "is an ill-composed entity which excludes most of the elements that allow it to exist" (p. 28). Pasteur, for instance, sets up a lab in which he can be the strongest: in the lab, the size of the phenomenon can be controlled, scaled, and translated (1988, p. 74).

INSCRIPTIONS

And that brings us back to the triangles in chapter 1. Like maps, like laboratories, these triangles cover territories (an increasing amount of territory in each iteration), amplifying some aspects of what they represent while reducing or eliminating others. They allow specific audiences to focus on what matters (or is taken to matter) in the phenomenon they represent. They scale the phenomenon (a moment of learning, a long-term collaboration, global climate change) into a form that can be simplified to its elements and analyzed. In fact, they work by simplifying, cutting the details that don't matter and the propositions that aren't acceptable.

But each cut detail is potentially a cut in alliances. Critics look at Engeström's triangle and declare that it has lost exactly what would have made it useful to them: Marxism (Elhammoumi, 2001), philosophy (Bakhurst, 2009), a unified concept (Blunden, 2010), signs (Miller, 2011), and on and on.

These triangles are *inscriptions*. As noted in chapter 1, Latour argues that two-dimensional inscriptions such as Engeström's triangles are potent ways to produce facts because they are mobile, immutable, flat, scalable, reproducible, recombinable, superimposable, incorporable into a written text, and geometric (1986, pp. 19–20). Having these qualities, these chains (Latour, 1987) or cascades (Latour, 1986) of re-representations allow us to create abstractions of what they represent, abstractions that retain only the most salient and comparable things. As Latour declares, "We write texts, we don't look through some window pane" (2006, p. 122). Scientists, he argues, can see phenomena only by looking at these inscriptions: "If scientists were looking at nature, at economies, at stars, at organs, they would not *see* anything. . . . Scientists start seeing something once they stop looking at nature and look exclusively and obsessively at prints and flat inscriptions" (1986, p. 16). That is, they must re-represent the *territory* (i.e., the phenomenon) as a flat, scaled, superimposable *map* (such as a triangle diagram; 2006, p. 133). Such re-representations allow us to "dominate" the phenomenon, scale it, compare it to others that have been similarly represented, and generate still more texts based on it. What maps take away in terms of detail, they give back in relationality.

We see something like this in the figures in chapter 1. Figure 1.1 shows an abstract model of individual mediation (illustrated by Vygotsky in concrete examples such as a knot in a handkerchief), while figure 1.2 shows an abstract model of a collective social system, which could be as small as a two-person collaboration or as large as an entire organization. These figures describe different spatial, temporal, and social scopes in which people learn and develop.

Such re-representations show up frequently in ANT studies. For instance, in their study of the Salk Institute, Latour and Woolgar (1979) demonstrate that the institute performs science by progressively re-representing concrete actions—the results of grinding up rats' brains, for example—as numbers, then tables summarizing numbers, and then graphs that are eventually published in articles. It is through such chains of re-representations, they say, that we *do* science and technology. Re-representing is a form of translation.

Re-representations can also include models, prototypes, and other technical innovations (Latour, 1996; Law, 2002). Furthermore, they are not simply the inventions of individuals. The graphs that Latour and Woolgar describe must be shareable with and recognizable by others. That is their whole purpose.

They must also undergo trials. And, critically, not all of these re-representations survive the trials to which they are put.

TRIALS

Lest inscriptions seem too much like writing fiction, they do have to withstand *trials* or tests by skeptical audiences who check the soundness of the cascade of re-representations (Latour, 1987, p. 75). Is this cascade of re-representations sound, or have the scientists missed something? Trials tend to take place at a specific time, under specific conditions, to test specific relationships (p. 78). Some of those trials are in the lab: a psychological theory is not very useful if it doesn't allow us to predict or explain anything. Thus, when Vygotsky & Luria (1930/1993) argue for the principle of mediation, they must demonstrate that this principle is evident in actual learning, so they point to Leontiev's experiments with memory. Science, even psychological science, should be able to withstand such trials. Through repeated trials, a claim, which is itself always potentially reversible, always testable, can be strengthened or weakened, made more or less objective—and more or less persuasive (Latour, 1987).

The Vygotsky Circle's experiments were, and are, vulnerable to such trials. Long after Vygotsky and Luria referred to Leontiev's "forbidden colors" experiment, Rene van der Veer tried to replicate it and got different results (2008). Van der Veer characterized elements of the original study's methodology as "a little naïve" (p. 28). Another of the other famous experiments associated with Vygotsky, the waiting room experiment (which he described vaguely in three different publications), appears to either be a thought experiment, a garbled version of a different experiment, or an account of an unpublished experiment (Sannino & Engeström, 2023). As Cole and Scribner argued in their introduction to Vygotsky's (1978) *Mind in Society*, Vygotsky "presents almost no raw data and summaries are quite general," lacking statistical tests and appearing to Western psychologists as pilot studies[1] rather than full-fledged studies. "In these writings Vygotsky was concerned with presenting the basic principles of his theory and method. He drew upon the very limited pool of empirical work available to him in order to illustrate and support these principles." Cole and Scribner attribute this lack to (1) Vygotsky's attempt to lay an entire base for a new Marxist psychology and (2) his limited time to do so, since he was dying of tuberculosis (p. 11). They note that Vygotsky's method was qualitative as well as quantitative, involved detailed description based on careful observation, and drew no barrier between laboratory and field research (p. 14). This approach was carried further by Luria, whose "romantic science" (Luria et al., 1979) was description-rich and statistics-poor, generating detailed case narratives (Luria, 1968, 1972; Luria & Udovich, 1956/1959). Luria, always a canny promoter, encouraged Oliver Sacks to take up this approach as well (Sacks, 1999, 2014).

Granted, small-scale qualitative studies are more difficult to replicate than closely controlled quantitative experiments, and they can be confounded by cultural and historical differences in study settings. What is striking here is not that replication failed, but that Vygotsky and colleagues staked so much on these small-scale, one-off studies rather than developing their robustness through either replicating them or drawing more broadly on similar research studies. Social scientists who work with similarly scaled studies typically mitigate the problems of nonreplication by comparing many different studies (e.g., Dumez, 2015). Similarly, the Vygotsky Circle could have conducted several similar studies and demonstrated a broader tendency across those studies. Put another way, Vygotsky's and his colleagues' experiments were not rigorously conducted by the standards of contemporary psychology or contemporary qualitative research. Consequently, they were quite vulnerable to trials such as van der Veer's replication study. They were often conducted in the field—in schools and rehabilitation facilities—and even when they used statistical methods, these tended to be quite rudimentary (e.g., Leont'ev & Zaporozhets, 1945/1960). Without strong empirical methodology—that is, strong self-inflicted trials—they were vulnerable to trials applied by outsiders. For instance, Luria's claims about his cross-cultural investigations in Uzbekistan were flatly contradicted by gestalt psychologist Kurt Koffka, who had witnessed the investigations and filed a separate report (in Luria, 1934; for attempts to replicate Luria's Uzbek studies, see Arunkumar et al., 2021; Scribner & Cole, 1981; Tulviste, 1991). We'll discuss this case in more detail in chapter 11. Similarly, Leontiev published claims about dermal sensing that amounted to parapsychology (Leontyev 1959/1981; see Leontiev, 2005;

for a contemporaneous discussion, see Zinchenko et al., 1974 and Maire & LeMoth, 1975) and are not seriously entertained today.

Without that sort of safeguarding, it's easy to overlook methodological shortcomings and focus on the results. That's arguably what happened in the studies described above. The exception was Luria's Uzbek research, which was genuinely replicated—and whose methodological problem was caught in the replication by Koffka. Stepping outside of CHAT, another obvious example is Lysenko's conversion of winter wheat into spring wheat, which he then generalized to embrace Lamarckian inheritance; Stalin's support prevented other scientists from rigorously testing these claims.

Still, the "romantic" small-scale qualitative studies performed by the Vygotsky Circle faced, and to an extent weathered, a lot of other trials. As Latour argues, a text is a "functional equivalent of a laboratory. It's a place for trials, experiments, and simulations" (2006, p. 148). And many of those trials took place in the conferences and journals of the day: as we've seen, Vygotsky and Luria's argument was challenged by Talenkin on ideological grounds, while the publication of Leontiev's book was delayed for similar reasons. As we'll see in the next part, Leontiev's many adjustments to the theory (see chapter 5) helped it to weather Stalinist trials (in both senses of the term). Surprisingly, given its empirically sordid past, the resulting translation successfully weathered other trials in the West, providing competition for a supposedly more rigorous American rival, information-processing cognitive psychology (chapter 6). Rather than scorpions stinging frogs for no reason, we'll see people struggling to preserve CHAT by translating it to meet different challenges.

THE TRANSLATION OF CHAT

Translation, then, is a materialist frame that has been used for understanding emergent rhetorical arguments in complex cross-specialty domains. It is often used to discuss how human and nonhuman actors perform interessement with each other, forming an alliance or relatively durable network of relations on which further agreement and actions can be built. And as I mentioned in the introduction, its landmark studies have been of the hard sciences and concrete technologies. These include studies of marine research (Callon, 1986a), biological laboratories (Latour & Woolgar, 1979), vaccine research (Latour, 1988), transportation systems (Callon, 1986b; Latour, 1996), aircraft (Law, 2002), bush pumps (de Laet & Mol, 2000), gazogenes (Akrich, 1993), telecommunications networks (Spinuzzi, 2008), and the development and commercialization of technological innovations (Miettinen, 1999; Akrich et al., 2002a, 2020b; Spinuzzi et al., 2016, 2018).

These examples are mainly in the so-called hard sciences and technology. But it is not just the hard sciences that are translated. In his later work, for instance, Latour applied a similar analysis to law (2010) and religion (2018), while Callon applied it to the travel industry (2002) and health science communication (Callon et al., 2009), and MacKenzie applied it to finance (2008). We also see some application to the methodology of participatory design (Spinuzzi, 2005a) and to technical communication (e.g., Swarts 2011). But we rarely see it applied to the social sciences proper. We certainly haven't seen a book-length study analogous to Latour's study of Pasteur (1988), in which translations of a social science framework are examined over decades.

But social science frameworks also emerge from translation processes, as we'll see, and understanding these translations can help us to better understand the frameworks emerging from them.

Translation emerged from the social sciences; let's turn the lens of translation back to the social sciences. Specifically, we'll examine the evolution of Engeström's CHAT, a specific theoretical and methodological framework that is often deployed in similar cases, and sometimes as a rival to translation (Spinuzzi, 2008; Kaptelinin & Nardi, 2006; Miettinen, 1999). We'll overview Engeström's CHAT in chapter 3, creating our own diagrams to understand it. Then we'll examine key translations in the rest of the book: points at which people redefined problems so they could position CHAT between a problem and a solution, thus gaining resources for their own research and sometimes saving their own careers.

In our translation analysis, we will attend to four elements: domain, project, translations, and trials.

DOMAIN

In what *domain* is CHAT deployed? That is, to what disciplines, fields, and pursuits is CHAT applied? As we'll see, CHAT is at various times applied to psychology, pedology, education, human-computer interaction, workplace learning, writing studies, organizational sociology, and public policy—among other domains. These domains have different needs, different associational networks, and CHAT had to be translated to address each.

PROJECT

What *project* did a key figure endorse when translating CHAT? That is, how did they articulate an endpoint within the domain, an objective that could secure the interest of others? As Latour argues, "By definition, a technological project is a fiction, since at the outset it does not exist, and there is no way it can exist yet because it is in the project phase" (1996, p. 23). Technologists are thus "novelists. With just one difference: their project—which is at first indistinguishable from a novel—will gradually veer in one direction or another. Either it will remain a project in the file drawers (and its text is often less amusing to read than that of a novel) or else it will be transformed into an object" (p. 24). In each chapter of the translation analysis, we examine an overarching project endorsed by a key figure, a project in which the key figure attempts to interest others.

TRANSLATIONS

How was CHAT *translated*—altered in its uptake—as it was applied to a domain in the service of a shared project? We consistently examine three categories of translations:

- **Empirical focus**: In its studies, what scope of human interaction did CHAT investigate? What methods did researchers draw upon to adequately examine that scope of interaction?
- **Mastery**: What were human beings portrayed as mastering within that scope? How was human agency portrayed?
- **Mediators**: What means were human beings portrayed as using in their attempts at mastery? What means were theorized and investigated?

TRIALS

Finally, to what trials was CHAT subjected? How were those trials related to domains, projects, and translations, and what were their results?

With this method established, let's now turn to CHAT, our research object.

3 CHAT'S GENERATIONS—OR STRANDS, OR WAVES?

The research object that we'll examine via translation is cultural-historical activity theory, or CHAT—specifically, Engeström's CHAT. As mentioned in chapter 1, I'll use "CHAT" as an umbrella term for the family of approaches tracing their lineage to Vygotsky and/or Leontiev, and I'll use "Engeström's CHAT" to designate the specific approach that Yrjö Engeström and his colleagues have developed.

CHAT is a framework for understanding learning and development. Engeström's CHAT is frequently used to study workplace education and learning in a variety of fields, to analyze and diagnose workplace problems, and to redesign work artifacts and processes—pursuits that Levant (2018) has characterized as "mundane." But more recently, and partly in response to critiques such as Levant's, Engeström and Sannino have expanded the scope of Engeström's CHAT to questions that are not mundane but, rather, are "radical and fateful" (Engeström & Sannino, 2021, p. 14), covering the interests of all inhabitants of the planet.

In part, it's this breadth of applications that makes CHAT an interesting object of a translation study, since it has had to interest a large number of actors to become so broadly applicable—especially considering how much CHAT has changed in order to offer this degree of flexibility.

But what *is* CHAT, and how has it changed? The answer is complicated. In this chapter, we'll first cover what we mean by CHAT. Then we'll map out three different metaphors of developmental change that have commonly been applied to CHAT: generations, waves, and strands.

WHAT IS CHAT?

The term "CHAT," or *cultural-historical activity theory*, is used as an umbrella term that covers both (1) the *cultural-historical* school that sprang up around Vygotsky's works as they became available in the West and (2) the *activity theory* school rooted in the later work of Vygotsky's colleague Leontiev (see Cole, 1996, p. 105).

Cole first heard of these two schools through his mentor Luria, who had worked closely with both Vygotsky and Leontiev. Luria, who in addition to being a top-notch neuropsychologist, was an energetic transnational broker and popularizer of scientific works (Metraux, 2021), had (around 1959, during de-Stalinization) begun to credit Vygotsky for the insights underlying his own neuropsychological work and had promoted Vygotsky's works in the West. One of those promotional efforts resulted in an English-language collection of Vygotsky's writings, edited by Cole and his collaborators, called *Mind in Society* (1978). This book set off a Vygotsky boom in the West and popularized the cultural-historical school. Chief among these insights was the claim in that collection's title: the individual mind is first and foremost social.

That same year, Leontiev's book *Activity, Consciousness and Personality* (Leont'ev 1975/1978) was also published in English. Although it did not have the same immediate impact as *Mind in Society*, it represented Leontiev's psychological theory of activity, in which consciousness and personality are developed and understood through the actions people take and the activities in which they jointly engage. Leontiev had worked closely with Vygotsky and Luria in the late 1920s and early 1930s; after Vygotsky's death in 1934, Leontiev adapted the theory of activity by combining Vygotsky's theory of mediation with Rubinshtein's category of activity. This category of activity was in turn based on Frederich Engels' (1971) account of how human beings created themselves through labor. In both Engels' and Leontiev's accounts, collective labor activity makes us human.

These two accounts share a lineage. In fact, Zinchenko (1995, p. 40) says that although they are two separate paradigms or strands of research, they should be considered the same school, with the same founders and the same basic assumptions. Between these two accounts, he says generously, there are vital, life-giving contradictions (p. 51). When we consider CHAT, we might consider a suture between the two, like this: *CH+AT*. This suture might heal

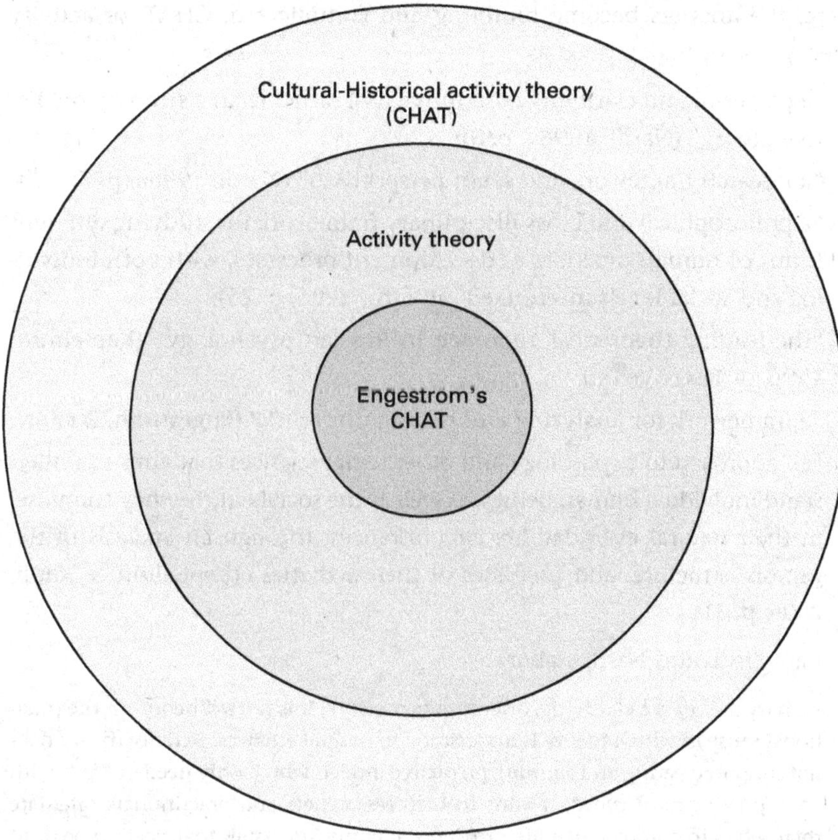

FIGURE 3.1
One way to conceptualize the relationships among CHAT, activity theory, and Engeström's CHAT

the wound—but then again, maybe not, as evocatively asserted in the combative title of one article: "Activity Theory Is a Dead End for Cultural-Historical Psychology" (Toomela, 2000).

At the same time, that suture enables a lot of slippage: sometimes people write about *CHAT* when they mean *activity theory* more specifically, and *activity theory* when they are more specifically talking about *Engeström's CHAT* and its emblematic triangle (see figure 3.1).

So far, this discussion gives us an idea of what the *letters* in CHAT stand for. But what *is* CHAT—a theory, a paradigm, a methodology, a framework?

Here, the answers become fumbling and complicated. CHAT, or activity theory, has been described as

- "a powerful and clarifying descriptive tool rather than a strongly predictive theory" (Nardi, 1996a, p. 3);
- "a research framework and set of perspectives" (Nardi, 1996a, p. 7);
- "a philosophical and cross-disciplinary framework for studying different forms of human practices as development processes, with both individual and social levels interlinked" (Kuutti, 1996, p. 25);
- "the leading theoretical approach in Russian psychology" (Kaptelinin, 1996, p. 103);
- "a framework for analyzing and redesigning work" (Engeström, 2000);
- "an approach in psychology and other social sciences that aims to understand individual human beings, as well as the social entities they compose, in their natural everyday life circumstances, through an analysis of the genesis, structure, and processes of their activities" (Kaptelinin & Nardi, 2006, p. 31).

Kaptelinin and Nardi elaborate:

> Activity theory is a clarifying, orienting framework. It is not a "theory" in the traditional sense in which theory is understood in natural sciences. Activity theory does not support creating and running predictive models which only need be "fed" with appropriate data. Instead, it aims to help researchers and practitioners orientate themselves in complex real-life problems, identify key issues that need to be dealt with, and direct the search for relevant evidence and suitable solutions. In other words, the key advantage of activity theory is to support researchers and practitioners in their own inquiry—for instance, by helping to ask right questions—rather than providing ready-made answers. (Kaptelinin & Nardi, 2012, pp. 6–7)

What is CHAT? A *framework*: a set of concepts, terms, and heuristics (such as Engeström's triangle) for understanding how people collectively address common labor objects. This framework has often been described by way of listing different principles (see table 3.1).

These sets of principles overlap quite a bit. Let's boil them down to three essential points: mind, mediators, and activity.

MIND IN, AND OF, SOCIETY

Cole and his coeditors named their 1978 English-language Vygotsky collection *Mind in Society*—an apt summary of this theme. CHAT in all of

Table 3.1
Principles of CHAT according to different authors

Source	Principles
Kaptelinin (1996)	• Unity of consciousness and action • Object-orientation • Hierarchical structure • Internalization and externalization • Mediation • Development
Kuutti (1996)	• Activities as basic units of analysis • History and development • Artifacts and mediation • Structure of an activity • Levels of an activity
Engeström (1999a)	Dichotomies in principles: • Psychic process versus object-related activity • Goal-directed action versus object-related activity • Instrumental tool-mediated production versus expressive sign-mediated communication • Relativism versus historicity • Internalization versus creation and externalization • Principle of explanation versus object of study
Kaptelinin & Nardi (2006)	Two main ideas: • Unity of consciousness and activity • Social nature of the human mind Seven basic principles: • Object-orientedness • Hierarchical structure of activity • Internalization-externalization • Mental processes versus external behavior • Interpsychological versus intrapsychological • Mediation • Development

its variations understands individual consciousness as being developed through that individual's interactions with others. For Vygotsky, such interactions begin at birth. For instance, an infant might reach vainly for an object that is then handed to her by a parent. Eventually the grasping (an attempt to control the environment directly) becomes pointing (a mediatory sign that helps to "control" the parent). The mediator not only helps to carry out an activity that would be difficult otherwise, but it changes the character of the activity. It qualitatively transforms it.

And in learning and taking up a culture's mediators, we become acculturated. In learning these physical tools and signs, we also learn what it is to be human in the networks of association in which we find ourselves operating. As Leontiev said, "one is not born a personality, one becomes a personality by socialisation and enculturation, by acquiring the habits, skills and methods of handling tools. Personality is a product of social activity" (quoted in Levitin, 1982, p. 123).

In the Soviet context, *social activity* meant *labor activity*. I use the term *labor* advisedly, but broadly. Emerging from the post-Revolutionary Soviet Union, CHAT is grounded in the philosophical tradition of Marx and Engels. Thus, it takes seriously the contention that the unique characteristics of human beings emerged from their cooperative activity: as Engels put it, labor "is the primary basic condition for all human existence, and this to such an extent that, in a sense, we have to say that labour created [humanity itself]" (1971, p. 279). This creation story was retold in various ways by Vygotsky and Luria (1930/1993) and Leontiev (Leontyev, 1959/2009b). Leontiev states that Marxist psychology "bases the psychology of the individual on the primacy of activity through which [humans] asserts [themselves] as personality"—production, or labor, rather than consumption (Levitin, 1982, p. 122). CHAT thus orients to ongoing, cyclical, collective efforts to achieve an objective.

Because of this view, CHAT is focused on the inherent sociality of consciousness and personality and, specifically, on how individuals and collectives develop over time. This inclination led to the denotation "cultural-historical," which was first deployed as a slur by critics (Razmyslov, 1934/2000; see Keiler, 2018) and then embraced by people working in the CHAT tradition.

MEDIATORS: PHYSICAL AND PSYCHOLOGICAL

Above, I mentioned *mediators*. Vygotsky (see figure 1.3a) distinguished between two sorts of mediators:

- **Physical tools**, which were oriented to transforming the physical world through labor
- **Psychological tools** (or signs), which were oriented to transforming or controlling one's own behavior

Unlike apes, who use physical tools to subjugate *nature*, human beings use psychological tools (signs) to control their *own behavior*—and "this constitutes the essence of the cultural development of man's behavior" (Vygotsky & Luria, 1930/1993, p. 77; cf. Vygotsky 1931/1997c, pp. 61–62 and Vygotsky 1930/1997d, p. 85). To illustrate the difference, Vygotsky and Luria quote Bucher's description of Borneo digging sticks, which are composed of a big stick and a little stick. The big stick is for digging: a technical *tool* oriented to labor. The little stick clicks with each stroke, creating a *sign* for regulating behavior: "This sound, being something like a work cry or a command, has the purpose of rhythmically organizing work" (p. 138).

It is the little stick that fascinated Vygotsky, the path by which a physical tool could become a sign (Vygotsky, 1929/1989). Figure 1.1 illustrates what he took to be the essential unit of analysis for studying this: the sign that someone selects to mediate between their stimulus and response, allowing them to control their own behavior through a cultural artifact. In a review of Vygotsky's works, Wertsch (2007) draws a further distinction between two sorts of sign mediation: explicit and implicit.

- In **explicit mediation**, individuals used physical artifacts (such as colored cards in Leontiev's forbidden colors experiment) as a second set of stimuli to control their own behavior.
- In **implicit mediation**, individuals instead used internalized representations such as inner speech.

But Vygotsky's associate Leontiev later collapsed this distinction between tool and sign. Shortly after Vygotsky's death, Leontiev (1937/2005a) characterized Vygotsky's focus on signs as idealist, charging that Vygotsky used the environment as a "magic trick," a construct that hid his idealistic views (p. 20). In contrast, Leontiev argued, word meaning develops through *productive activity*, that is, labor (p. 24). "Thus, Vygotsky's proposition that consciousness is a product of the child's verbal communication under conditions of his activity and in relation to the material reality that surrounds him must be turned around: the consciousness of a child is a product of [their] human activity in relation to objective reality, taking place under conditions of language and under conditions of verbal communication" (p. 25). This productive activity is carried on via social relations and mediated by tools. Thus, in Leontiev's view, the separation between tool and sign is unnecessary and idealist; it locates meaning in the wrong place.

Later, building on the work of Vygotsky, Luria, and Leontiev, Cole (1996) treated both tools and signs under the heading of *cultural artifacts*, which "are simultaneously ideal and material . . . combin[ing] the properties of tools and symbols" (p. 144). Cole argues that artifacts

- "are the fundamental constituents of culture";
- "do not exist in isolation as elements of culture";
- "exist as such only in relation to 'something else' variously referred to as a situation, context, or activity" (p. 144).

Such artifacts are physical, but they are often designed to work as a system whose physical characteristics are crafted to support semiotic ones (see Schmandt-Besserat & Erard, 2008; Olson, 2016).

But, as we'll see below, other mediators have been identified beyond artifacts.

THE STRUCTURE OF ACTIVITY

Kaptelinin and Nardi (2012) note that while "Leontiev's approach is predominantly concerned with activities of individual human beings," Leontiev "does not systematically explore the structure and development of collective activities and does not present a conceptual model of collective activity" (p. 33). Such a model was later proposed by Engeström, based on his reading of Leontiev. Engeström's (1987/2014) famous triangle (figure 1.2) presents a model for identifying, analyzing, and elaborating on the labor activities that Leontiev had described. In this figure, the *activity system*, Engeström identifies three points (subject, object, and community) and three kinds of mediators sitting between them:

- **Tools** (or instruments) mediating between subject and object
- **Division of labor** mediating between object and community
- **Rules** mediating between subject and community

In this structure, a given activity is anchored by the *object*, which Kaptelinin calls "the sense–maker," because it "gives meaning to and determines values of various entities and phenomena" (Kaptelinin, 2005, p. 5). The object is both *objective* (raw materials) and *projective* (the use-value that an actor seeks to achieve by transforming it). This object is cyclically transformed to reach some envisioned outcome. In identifying an object, an activity theorist identifies the other parts of the activity system that are arrayed to transform it (e.g., Spinuzzi, 2018). This activity system functions as a unit of analysis that delimits the phenomenon under study. In more colloquial terms, the object is the "text" to which the rest of the activity system is the "context," the principle by which the analysis is delimited.

The object is transformed by a *subject* or *actor* (the terms are sometimes used interchangeably) through a *tool* or *instrument*. It is transformed according to explicit and implicit *rules* that the subject shares with a *community*. And that community contributes to the transformation via a *division of labor*. Critically, the object is transformed cyclically, meaning that the activity system describes repeated transformations, and as the activity system itself develops over time, so does the object.

For instance, suppose that the activity under study is farming: a farmer labors to cyclically transform their field "from brute earth to crops of grain in the process of work activity" (Engeström & Escalante, 1996, p. 360). We can identify instruments (such as hoes or Borneo digging sticks), rules (such as when to plant and when to harvest), and a division of labor (such as who plants and who harvests) that the farmer uses. As the farmer continues to labor season after season, they develop expertise—but they might also change motivations from subsistence to profit, they might pick up new instruments, or they might develop (or be forced to conform to) new rules. The activity system "incessantly reconstructs itself" (Engeström, 1992, p. 12).

These mediators are typically given within the activity under study—that is, these cultural artifacts are also cultural inheritances. The farmer's instruments, rules, and division of labor come from previous farmers who have developed solutions in their own activities, and sometimes from other activities (for instance, a newly manufactured piece of farm equipment might be introduced into the activity, revolutionizing it). Further, the farmer might develop or customize these mediators as they are integrated into the activity.

CHAT involves more, much more, than these three points. We'll explore other aspects in the rest of the book. But these points should give us enough grounding to get the discussion started. Specifically, they will help us to analyze changes in three specific aspects of CHAT, aspects to which we will return in part II:

- **Empirical focus**: What is investigated in a CHAT study? What are the bounds of a CHAT investigation? What is its unit of analysis?
- **Mastery**: What is being mastered, developed, or transcended?

- **Mediators**: What cultural artifacts are being used in the investigated frame, and for what?

Part II is about how CHAT has changed. Before we embark on this detailed examination of change, let's see how others have told the story.

METAPHORS OF CHANGE IN CHAT

When CHAT history is told, it is typically told with one of three competing metaphors: generations, strands, and waves. All three metaphors tell a story of development—which makes sense, CHAT being focused on learning and development.

METAPHOR 1: GENERATIONS (THE STORY OF ENGESTRÖM'S CHAT)

In this metaphor, proffered by Yrjö Engeström in a number of publications (e.g., Engeström, 1996b; Engeström & Sannino, 2021), CHAT has developed in *generations*. Each generation faces a different problem, which it can only address by dialectically transforming itself. Engeström's story ascribes coherence across generations or traditions, demonstrating how each generation is built on (an interpretation of) a previous generation (table 3.2).

FIRST GENERATION This generation was centered around Vygotsky's insight into mediation, illustrated in Vygotsky's triangular model of stimulus, response, and mediating artifact (figure 1.3c); Engeström adds that the model is "commonly expressed as the triad of subject, object, and

Table 3.2
The four generations of CHAT according to Engeström

Generation	Features
First generation (Vygotsky)	Triangular model of mediation; individual focus
Second generation (Leontiev)	Mediation understood within a historically evolving division of labor, differentiating between individual action and collective activity
Third generation	Triangular model of activity; collective focus; activities interacting in networks; dialogue; contradictions
Fourth generation	Coalescing cycles of learning in heterogeneous coalitions; alternatives to capitalism

mediating artifact" (1996b, p. 132). Inserting a cultural artifact into this model "was revolutionary in that the basic unit of analysis now overcame the split between the Cartesian individual and the untouchable societal structure. . . . Objects became cultural entities and the object orientedness of action became the key to understanding human psyche" (Engeström, 1987/2014, p. xiv). Yet, Engeström says, in this generation the focus was limited to the individual.

Engeström does not make clear who reformulated the triangle as subject-object-mediator, leading Miller (2011, p. 220) to make the incendiary claim that this reformulation was backfilled from Engeström's own triangle (figure 1.3d). However, in one of his unpublished 1929 papers, Vygotsky (1929/1989) presents both triangles side by side: The first represents mediation of labor with a tool, directed outward, while in the second, the tool becomes a sign, "placed between me and my memory" (p. 61; see figure 1.3a). That is, Vygotsky is not just delineating between tools and signs; he is showing how one leads to another.

SECOND GENERATION The limitation of individual focus was "overcome by the second generation, led and inspired by Leont'ev's work" (Engeström, 1987/2014, p. xiv). Leontiev "showed how [the] historically evolving division of labor had brought about the crucial differentiation between an individual action and a collective activity" (pp. xiv–xv). Yet, Engeström says, "the relationship between object-oriented production and communicative exchange between people remained somewhat unclear in Leont'ev's work," and Vygotsky's triangle diagram was never graphically expanded in this work (p. xv). Engeström graphically expands the triangle to address these relationships (figure 1.3d), and in doing so, he adds the elements of community and rules. Beyond that, in the Soviet Union, activity theorists mostly restricted their studies to "play and learning among children" (as well as rehabilitation and "defectology," or special needs education), and "contradictions of activity remained an extremely touchy issue" (p. xv). In the 1970s, he recounts, radical researchers in the West picked up the concept and applied it to other domains, including work, and philosopher Evald Ilyenkov's writings on contradictions became incorporated (p. xv).

THIRD GENERATION Here, Engeström drops the human actors in his sentences, but the described agenda is the one that he himself defined:

> The third generation of activity theory is developing conceptual tools to understand networks of interacting activity systems, dialogue, and multiple perspectives and voices. In this mode of research, the basic model is expanded to include minimally two interacting activity systems. This move toward networks of activities, while still in an embryonic form, was anticipated in the original text of *Learning by Expanding*. (Engeström, 1987/2014, p. xv)

Engeström is referring to the concept of the *activity network*, in which two or more activity systems interact (figure 1.3e and f). For instance, a patient and a doctor may be involved in different activities but meet during a doctor-patient visit in which they share the same object (the patient's well-being). Or one activity might supply the instruments for another. For instance, a farm equipment manufacturer might supply our farmer with a new tractor, while an agricultural conglomerate might supply the farmer with genetically modified seed. By diagramming such activity networks, an activity theorist can conceptualize how historically different activities interact—a critical step if one's goal is to model organizations and their interactions.

To describe how such activities change, Engeström uses the concept of *contradictions*, specifically drawing on Soviet philosopher Evald Ilyenkov's writings. For Engeström, "Contradictions are historically accumulating structural tensions within and between activity systems" (2001, p. 137). We'll discuss contradictions in more detail in chapter 6.

In putting these concepts into practice, early on, Engeström and his collaborators relied on small-scale *field* studies using observations and interviews as well as conversation analysis. But they also began using—and later moved almost exclusively to—*interventionist* studies, using a technique called *change laboratory*, in which researchers coordinated participants' deliberations about their own activities. Change labs taught these participants enough of the concepts, vocabulary, and inscriptions of third-generation CHAT to support these deliberations. One critical concept was the cycle of expansive learning (CEL), which describes how organizations learn together. This concept mapped out a cycle of questioning, analyzing, modeling a solution, examining the solution, implementing the new model, reflecting on the process, and consolidating and generalizing the new practice (Engeström & Sannino, 2010, p. 8). The CEL describes (and prescribes) how change labs are run.

FOURTH GENERATION In a much later article, published thirty-four years after *Learning by Expanding*, Engeström and Sannino argue that it is time to consider a fourth generation of activity theory:

> To us, the main challenge that requires a fourth- generation activity theory is the radical and fateful transformation in the *objects* of human activity. In the current phase of capitalist globalization, such interconnected objects as poverty, climate change, and pandemics cannot anymore be treated as isolated issues to be brought under control by technical means; they influence and pervade the objects of innumerable activities and call for radical revisioning of the ways our societies and lives are organized. (Engeström & Sannino, 2021, p. 14)

Engeström and Sannino have specified that "the challenge of [the] fourth generation of activity theory is alternatives to capitalism. How you build sustainable viable resilient alternatives to capitalism especially understood as the neoliberal global regime" (quoted in Ploettner & Tresseras, 2016, p. 93). Their approach to doing so is by expanding the interventionist methodology of change labs and the concept of the cycle of expansive learning to explore interactions across multiple institutions. One result is Engeström and Sannino's figure of intersecting loops, depicting the "coalescing cycles of expansive learning" across institutions (Engeström & Sannino, 2021, p. 15).

For Engeström—and for other commentators on the development of activity theory (e.g., Kaptelinin & Nardi, 2006; Spinuzzi, 2011; Yamazumi, 2008)—the key spokespeople of this tradition are Vygotsky (first generation) and Leontiev (second generation). For the third generation, we can point to Engeström himself as a key spokesperson. For the fourth generation, Engeström and Sannino have been at the forefront.

This story is told with inscriptions, including triangles, which increase in complication over time. As noted in chapter 2, Latour argues that inscriptions stay roughly the same size, scaling the phenomenon they represent to make it comparable to other representations. Certainly we see that in this parade of inscriptions. All are roughly the same, but they claim more and more expansive territory: spatially, temporally, and socially. In the generations account, CHAT keeps learning by expanding, depicting vaster and vaster contexts, progressing from a knotted handkerchief to the entire world.

However, this "generations" approach has been criticized for its narrative of progression, one that encourages us to understand CHAT as evolving *linearly*: the theory encounters a linear series of problems and is serially reformulated to address those problems. This narrative has the effect of legitimizing specific problems and changes—those problems that most concern the narrator—while others are ignored. Yet others argue that activity theorists have discerned and addressed different problems, leading not to a linear set of developments but to an explosion of developments in various directions. As Stetsenko argues,

> CHAT today is a diverse framework consisting of various currents and trends—it is not easy to break them up by either geographic location or generation (my understanding is that even within Finland there is more than one CHAT-related, Vygotsky-inspired research direction). It is, in fact, one of CHAT's strengths that it has evolved into and spawned a heterogenous, diverse collectives [sic] of scholars working all over the world, in various disciplines from education and work studies to literary studies, research on digital technologies, human-computer interaction, psychology, and the study of language, among others. (2021, p. 39)

More pointedly, Bakhurst charges that "the account of the three generations of activity theory is something of a 'just so' story. The problem is not only the omission of numerous Russian thinkers who contributed to the Soviet school . . . the point is rather that, within Russian philosophy and psychology, the concept of activity was always seen as problematic and open to multiple interpretations" (2009, p. 201). Bakhurst is correct to a point, but it is also true that the Soviets were not eclectic in their approaches. Writing in 1975, McLeish noted that Soviet psychology "rejects eclecticism as a sign either of intellectual incompetence or of an intellectual compromise based ultimately on ideological self-interest" (1975, p. 65). Lest we think this is a case of Cold War overgeneralization, Leontiev said basically the same thing in an interview: "Soviet psychology has rejected the path of methodological pluralism of which Western psychology is so proud. I think this is false pride, because the old adage, 'many approaches mean no approach' has a lot of truth to it" (quoted in Levitin, 1982, p. 112). To support his approach, Leontiev diligently leveraged his own "just-so" story, the Engels-inspired story of how humanity created itself through labor (a story that Blunden calls a "Stalinist fairy tale"; 2010, p. 218). That is, Engeström's just-so story is part of a longer tradition of using stories to create coherence, coherence that is periodically counterbalanced by heterogeneous efforts such as the ones Stetsenko mentions.

The metaphor creates that coherence: like an individual tracing their patrimony, this account looks back at the development of this specific genealogy. When we read about Jesus' genealogy in Matthew 1:2—

Abraham was the father of Isaac,
Isaac the father of Jacob,
Jacob the father of Judah and his brothers . . .

—the genealogy does not tell us about Jacon's brother Esau or Isaac's brother Ishmael or the other branches of the family, because they are not important for establishing Jesus' lineage. (Judah's brothers *are* important because they represent the twelve Tribes of Israel.) Yes, like this genealogy, Engeström's generations metaphor is a "just so" story, one that looks back and sees a straight line of development. Its fidelity is to its lineage, with patriarchs (Vygotsky Leontiev, Engeström) developing the theory in response to trials. But Bakhurst, Stetsenko, and others see themselves being cut out of the story just like Esau and Ishmael: disinherited. The metaphor doesn't adequately represent CHAT's development. So let's turn to a second metaphor that promises to embrace these other developments.

METAPHOR 2: STRANDS

"Strands" is a popular and somewhat unexamined metaphor that is often applied to theory. Alex Levant uses it in his ecumenical article "Two, Three, Many Strands of Activity Theory!" (2018). Picking up on Bakhurst's (2009) assertion that activity theory has a philosophical strand and an organizational strand, Levant declares,

> I see activity theory as a broad tent, a theoretical tradition in the making. There are common origins and very different directions and developments from the perspective of which the origins appear to diverge as well. I am quite impressed with even the most mundane applications of CHAT; however, in agreement with Bakhurst, I see the strength of the tradition elsewhere. Certainly, the lines of development of activity theory in the Soviet Union had broader societal concerns, and they developed a body of thought that appears abridged in the triangle model. What I find most interesting about Soviet activity theory is its diversity and its orientation on human emancipation. (p. 107)

Levant concludes, "In this spirit of struggle and diversity, I call for two, three, many strands of activity theory" (p. 107). Figure 3.2 shows some of these strands.

Here, Levant acts as an intercessor, arguing for heterogeneity in CHAT, just as Stetsenko (2021), Sawchuk & Stetsenko (2008), and Lektorsky (1999) do. In the strands metaphor, different scholars take on different research objects, or take up CHAT in different disciplines, resulting in implementations that are both internally coherent and, because of their common starting point,

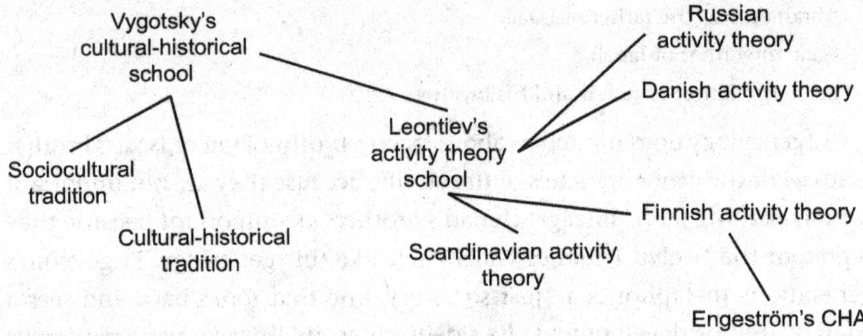

FIGURE 3.2
Some strands of CHAT

somewhat compatible with other implementations. These are "multiple readings" (Van der Veer, 2008). For instance, as we'll see in chapter 6, Engeström and Bødker worked independently to apply Leontiev's activity theory to learning theories in their own fields. They developed different parts of Leontiev's legacy, and when they discovered each others' work, they cited each other (to an extent). Similarly, other strands of activity theory, cultural-historical theory, and CHAT have been picked up, developed separately, and then sometimes woven together by authors such as Bruner, Cole, Kaptelinin, Lave, Miettinen, Nardi, Säljö, Scribner, Wertsch, and many others.

In another example of the strands metaphor, Toomela (2008) identifies two strands of Vygotskian research: the sociocultural strand (represented by Matusov) and the cultural-historical strand. Although he identifies deficiencies in the sociocultural strand, specifically its disregard of cognition, he argues, "I suggest that the sociocultural and cultural-historical schools should not be opposed. Both of them, taken separately, have limitations that can be overcome by seeing them as representing two complementary levels of analysis of the same phenomenon that psychologists try to understand, the mind" (p. 58). (Toomela does not extend this level of grace to activity theory.) And in still another example, Zinchenko sees cultural-historical psychology and activity theory as comprising "two strands of research, but not two schools" (1995, p. 40), between which are vital, life-giving contradictions (p. 51). Finally, Hopwood (2022) argues for thickening connections across CHAT strands.

The strands metaphor is ecumenical: it recognizes that there are other lineages, lineages that are arguably internally coherent and thus legitimate or at least usable (although perhaps heterodox) in relation to their chosen objectives. Yes, a given author might regard their strand as more significant, theoretically rigorous, or faithful, but others are acceptable and they may develop ideas that can be woven back into the tapestry of CHAT. Here, the fidelity that matters is the fidelity within each strand; if that's achieved, we can live and let live, and perhaps something good may come from it. This detente across strands is what allows some of the landmark collections on CHAT to exist (e.g., Brown et al., 1999; Nardi, 1996b; Daniels et al., 2010; Engeström & Middleton, 1996; Sannino et al., 2009).

Yet this ecumenical tolerance has its limits, as demonstrated in Bakhurst's, Levant's, and Toomela's arguments. The three draw their lines differently, but ultimately some strands are regarded as more legitimate than others. Further, the strands metaphor assumes coherence that is not consistently practiced even within a "strand."

METAPHOR 3: WAVES

A third metaphor used in discussions of CHAT is that of *waves*. This metaphor is mainly used for discussing how Vygotsky's work was taken up: in critiques of hagiography, in additional Russian publications of source materials, and in English translations. Specifically, commentators using this metaphor tend to emphasize how the work was initially taken up *poorly* and how subsequent work has *dispelled* (or should dispel) poor uptakes.

The waves metaphor involves accretion, with subsequent waves depositing texts that overlay the previous ones, making interpretation progressively more well founded. Over time, more of the original source materials are published, allowing us to better and more clearly understand the work of previous years. Related, later versions of sources are often regarded as more accurate than earlier ones. Often, the poor quality and questionable interpretations of earlier work is blamed on incompetence, intrigue, or both.

IN CRITIQUES OF HAGIOGRAPHY This work has been more recently associated with Vygotsky "myth busting" and the "revisionist revolution in Vygotsky studies" (e.g., Yasnitsky & Van der Veer, 2016; Yasnitsky, 2018a, 2019, 2020a, 2020b). Authors in this vein tend to focus on how Vygotsky, in particular, was elevated long after his death, with different accomplishments—and views—attributed to him rather than to his sources or other members of his circle. For instance, Yasnitsky begins his biography of Vygotsky in this way: "This story is about a genius. So they say. But the person did not become [a] genius until after his death" (2018a, p. xi). Elsewhere, he notes that the Soviet Union tended to create hagiographies in which a small number of great figures, characterized as good or evil, played the roles of martyrs or victims (2016a, p. 24).

In critiques such as these, authors attempt to delineate the actual sources of ideas attributed to Vygotsky, such as the zone of proximal development (Valsiner & Van der Veer, 2014); identify instances in which a colleague's work was attributed to Vygotsky, such as the concept of double stimulation (e.g., McLeish, 1975; see Sakharov, 1994); and describe how Vygotsky's work was promoted in the West, beginning twenty-five years after his death, partly in order to bolster the international and Russian reputations of his colleagues Luria and Leontiev (Metraux, 2021). They also note how Vygotsky's colleague Leontiev, in particular, recast Vygotsky as a predecessor of his own work, a claim that was not seriously contested until after Leontiev's death (Kozulin, 1999, pp. 253–254).

IN ADDITIONAL RUSSIAN PUBLICATIONS This work focuses on the question of what Vygotsky actually wrote and published and when. This question may seem straightforward, but the waters have been muddied considerably. For instance, Vygotsky wrote several pieces around 1931 that, in 1960, were assembled and published by editors as a five-chapter book: *The History of the Development of Higher Mental Functions*. Much later, in Vygotsky's *Collected Works* (in Russian: 1982–1984, in English: 1997c), other materials were appended to it, extending it to a fifteen-chapter book. A second example is "Tool and Symbol in Child Development" (Vygotsky & Luria, 1930/1994), which is sometimes attributed to Vygotsky and sometimes to Vygotsky and Luria. This work has been called a "benign forgery" with eccentricities that possibly result from translating the English manuscript to Russian and then back to English (Van der Veer & Yasnitsky, 2016a). As a third example, Zavershneva and Van der Veer (2018) have recently published the contents of some of Vygotsky's private notebooks, shedding light on Vygotsky's influences and transitions. As waves of publications by Vygotsky and his collaborators are exhumed, rediscovered, rescrutinized, and produced, we (presumably) know more about his thought process—and about the intrigues and disagreements that shaped it.

One major reason that these publications have trickled out over such a long period is the fact that Vygotsky's works were associated with pedology, the interdisciplinary study of the child. Pedology was essentially banned as a field of study in 1936, shortly after Vygotsky's death; in fact, the Pedology Decree is sometimes considered the beginning of the Great Terror. Thus, Vygotsky's works were seldom read or cited in the remainder of Stalin's lifetime, and indeed, the publication of psychology research in the USSR nearly ground to a halt. Shortly after Stalin's death in 1953, the floodgates opened. In 1956, Luria published his twin study (Luria & Udovich, 1956/1959) and Vygotsky's family published an early collection of his works. In 1959, Leontiev published *Problems of the Development of Mind* (Leontyev, 1959/1981), and he and Luria began heavily promoting Vygotsky to their Western colleagues (Metraux, 2021). In 1960, various Vygotsky pieces from around 1931 were assembled into *The History of the Development of Higher Mental Functions*. As Vygotsky was methodically promoted in the West, his status in the USSR rose as well, yielding other publications of previously ignored, suppressed, or unpublished materials. And Vygotsky's higher profile in turn meant that his colleagues were motivated to attribute accomplishments to him as well as to portray him as a forerunner whose actual arc was completed by those associates (e.g., Brozek, 1972).

IN ENGLISH TRANSLATIONS In perhaps the paradigmatic example of the wave metaphor, Jaan Valsiner (1988) notes that Vygotsky's work underwent three waves of translations (figure 3.3).

- **First wave**, 1920–1939: All the basic aspects of Vygotsky's cultural-historical approach were available in English (Valsiner, 1988, p. 155), but the uptake was thin.
- **Second wave**, 1961–1981: Cognitive interests had been reestablished in Western psychology, and Vygotsky's works gained interest under that interpretation. (Luria and Leontiev had also begun heavily promoting Vygotsky in the West in 1959.) These English translations included two translations of Vygotsky's 1934 *Thought and Language* (1962; 1963) as well as 1978's *Mind in Society*, a "potpourri of Vygotsky's ideas, collected together from different sources and linked in ways judged to fit the 'recipient' culture better than the original" (Valsiner, 1988, p. 155). Importantly,

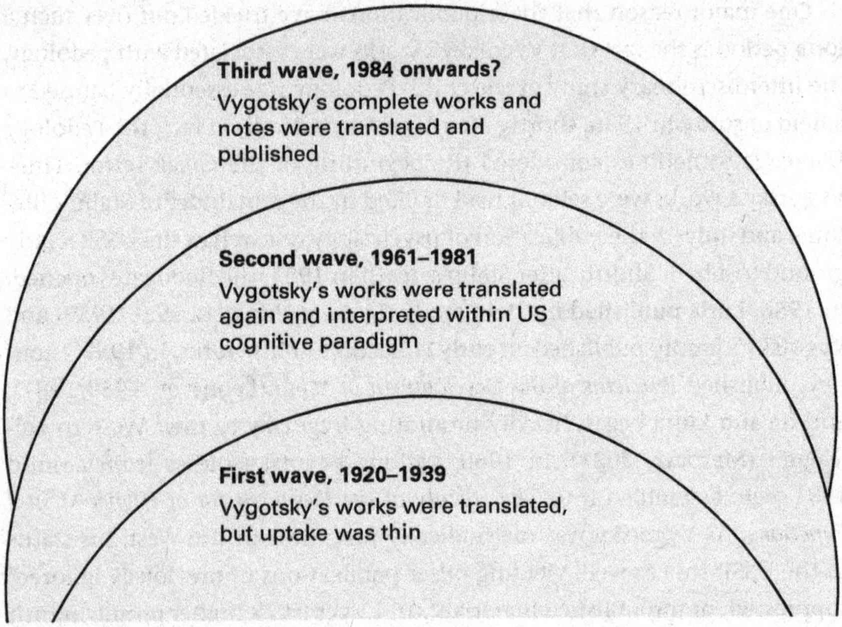

FIGURE 3.3
Valsiner's three waves of translations

part of the alteration for the recipient culture involved cutting cites to Marx and Engels. Valsiner presents a table showing citations to Vygotsky in the West, demonstrating that most of the citations are to the second-wave translations (p. 160).

- **Third wave**, 1984– : Fueled by the great uptake of *Mind in Society* in the West, many more of Vygotsky works were translated. Since Valsiner wrote this account in 1988, even more of Vygotsky's works have been published in English, including Vygotsky's *Collected Works* in 1997–1999, his notebooks (Zavershneva & Van der Veer, 2018), and various letters.

Whereas the strands metaphor is the most ecumenical, the waves metaphor is the least so. It is all about fidelity to what *really* happened in the past, to the source material—typically Vygotsky's works. What did Vygotsky *really* mean, believe, and say? How have the past waves misrepresented him? How did he have to guard his words in Stalinist Russia, and what would he have said, who would he have quoted, if he could? Those who use the waves metaphor tend to see earlier waves as either incompetently *interpreted* (adjusted for the recipient culture in ways that threaten the integrity of the source material) or deliberately *garbled* (hidden or kept quiet in the past due to the intrigue in Stalinist Russia). The more we know about the source materials—the more we see Vygotsky's personal notebooks or his correspondence with his circle, for instance—the stronger the fidelity. Thus, some of the people who were invited into the discussion via the strands metaphor have been once again disinvited, disinherited, seen as pretenders with spurious claims to the tradition. For the waves metaphor, the source material cannot and should not change.

But, of course, the problem with fidelity to the source material is that some of its arguments no longer have currency. Indeed, some of Vygotsky's arguments, such as his devotion to bringing about the New Soviet Human (chapter 4) or the hard-and-fast distinction between higher and lower mental functions, may seem outlandish to us. Vygotsky himself (as noted in chapter 2) changed some of his positions over his life as well, so keeping absolute fidelity to his works seems questionable.

All three of these metaphors focus on development and fidelity to their lines of development. The generations metaphor traces development from simple to complex and from smaller to greater areas of influence. The strands metaphor traces growth in the set of approaches that collectively make up CHAT. The waves metaphor traces growth in the understanding of unchanging source material, some of which (like Vygotsky's notebooks; Zavershneva & Van der Veer, 2018) perhaps still waits to be discovered.

This common focus on development should not be surprising, since all three metaphors are grounded in dialectic. Dialectic assumes that time's arrow goes in one direction, toward a more complexly mediated unity (Wegerif, 2008, p. 350). Yet these metaphors represent different *lines* of development, and the accounts compete to elevate one line above the others. This is why these accounts point out each others' infidelities. And it is also why CHAT is poorly equipped to analyze its own history: its analyses start with development and assume a dialectic. Its frames tend to be either an apolitical, administrative understanding of development (via Leontiev) or a highly political Marxist grand narrative.

ANOTHER METAPHOR: TRANSLATIONS OF CHAT

As discussed in chapter 2, we'll use a different metaphor. In the rest of this book, we'll think of CHAT as an innovation that, like other innovations, doesn't get diffused but, rather, gets translated. Like a vaccine, a transportation system, or a fighter plane, CHAT can be seen as a material innovation that must be transformed to be transported. This transportation is most visible in the uptake of its triangle diagrams (figure 1.3), but we can discern it in other ways as well.

Rather than trying to trace all the many changes and variations related to CHAT—an impossible task—I instead focus on one specific and relatively coherent version, Engeström's CHAT. How did it become the way it is? What translations led to it?

In the rest of this book, I respectfully but critically engage with the ideas of Vygotsky, Leontiev, Engeström, and others in CHAT's lineage and global community of practice. But I engage with these ideas via the sociology of translation. To travel from the Soviet Russia of the 1930s to the Finland of the 1980s and the United States of the 2020s, to be taken up in different networks of associations, CHAT had to be translated for those networks. One dimension of that transformation was in its disciplinary orientation: from Marxist psychology to an interventionist design approach to organizational sociology to (most recently) a method for implementing public policy. These changes are visible in the arguments advanced for CHAT, the cases to which it is applied, and the concepts that have been developed for it. Across the parts, I identify these moments of translation to better understand CHAT's transportation to different networks of association, including national boundaries and ideologies, domains, and empirical methods. Of special interest are changes in CHAT's empirical focus, mastery, and mediators at each point. This work is a continuation of my earlier book-length contrast of activity theory and actor-network theory (Spinuzzi, 2008).

As we saw in chapter 2, translations usually happen via many small deviations, deviations that are so minor that they are often hard to detect. For that reason, in part II, I take stock of (1) more significant translations that (2) are associated with and attributed to specific figures (3) during specific spans of time. I do this for convenience: it's a lot easier for us to discuss "Vygotsky's CHAT" or "Leontiev's CHAT" by summing up the translations associated with that person than it is to catalog the many granular deviations that had to happen. These iterations developed to address different needs over time rather than in a thunderbolt. When I discuss these different translations of CHAT, then, I am not referring to purified forms—an idea that is antithetical to the sociology of translation—but, rather, summarizing changes associated with specific figures.

This history of translations helps us to understand why contemporary CHAT is the way it is—why it has the strengths and limitations that it does. In so doing, it provides a critical foundation for understanding how CHAT treats acting with technology.

II TRANSLATIONS OF CHAT

4 VYGOTSKY'S CHAT: MARXIST PSYCHOLOGY BEFORE THE GREAT TERROR

What we now know as CHAT traces back to the work of Vygotsky, and specifically, the notion of mediation from his instrumentalist period. Vygotsky's concern was not workplaces but, rather, a fundamental explanation for how people learn and develop. In a passage illustrated by Vygotsky's triangular figure (figure 1.1), Yrjö Engeström notes that mediation

> was crystallized in Vygotsky's (1978, p. 40) famous triangular model of "a complex, mediated act" . . . which is commonly expressed as the triad of subject, object, and mediating artifact. . . .
>
> The insertion of cultural artifacts into human actions was revolutionary in that the basic unit of analysis now overcame the split between the Cartesian individual and the untouchable societal structure. The individual could no longer be understood without his or her cultural means; and the society could no longer be understood without the agency of individuals who use and produce artifacts. This meant that objects ceased to be just raw material for the formation of the subject as they were for Piaget. Objects became cultural entities and the object-orientedness of action became the key to understanding the human psyche. (Engeström, 1996b, p. 132)

Yet Engeström charges that although Vygotsky's work understood the individual as thoroughly social, it still took as its unit of analysis the individual's mediated action rather than a community's ongoing activity.

To better understand Vygotsky's CHAT, let's think about it in terms of domain, project, translations, and trials.

- **Domain:** The domain that Vygotsky addressed was psychology. Vygotsky and his circle attempted to translate psychology's understanding of fundamental aspects of human psychology (such as memory, perception, attention, and personality) into a historical materialist framework. To do so, they borrowed heavily from *predecessors in psychology* such as Pierre Janet, James Mark Baldwin, the American pragmatists, and George Herbert Mead (Valsiner & Van der Veer, 2000); *contemporaries in psychology* such as Piaget, Buhler, Kohler, Koffka, and Lewin (Zavershneva & Van der Veer, 2018); *Russian social scientists* such as Basov, Potebnya, and Shpet; and the *sociologist* Emile Durkheim (Van der Veer & Valsiner, 1991). These insights were taken up and translated through the insights of Marx and Engels to develop what Vygotsky characterized as a Marxist psychology—although, as we will see, other Soviets resisted that interpretation during the 1930s.

- **Project**: Vygotsky's project—what he was trying to achieve—was the New Soviet Human,[1] a human being whose potential had been unlocked by the transition from capitalism to socialism. This overarching project, which we'll discuss below, drove Vygotsky's attempt to develop a Marxist psychology suitable for producing the New Human.
- **Translations**: To produce a Marxist psychology suitable for bringing forth the New Human, Vygotsky translated antecedent psychologies in several ways. We'll specifically examine translations in empirical focus, mastery, and mediators. The sum of these translations was a new settlement, a cultural-historical psychology.
- **Trials**: Finally, this settlement had to weather trials. Unfortunately, as we'll see at the end of the chapter, Vygotsky's Marxist psychology did not weather these trials well, requiring the further translations discussed in chapter 5.

In this chapter, I first provide a historical account of how Vygotsky became involved in psychology and how he developed his psychology throughout his life. Next, I examine Vygotsky's psychology in terms of the three translations mentioned—empirical focus, mastery, and mediation—and how these translations led Vygotsky to address his project of achieving the New Human. I conclude by identifying trials that tested Vygotsky's Marxist psychology. After Vygotsky's 1934 death, the USSR descended into the *Great Terror* (1936–1938): Stalin's campaign of purges that solidified his power over the Communist Party and the USSR. In fact, the Great Terror began with the infamous 1936 Pedology Decree, which made Vygotsky's work difficult to cite until after Stalin's death in 1953. The resulting tensions would lead to a translation to Leontiev's Marxist psychology: activity theory. Any chapter-length review of Vygotsky's thought would be too short, but I hope to cover enough of the basics to understand how Vygotsky and his colleagues translated their antecedents into Vygotsky's cultural-historical theory.

Let's start with a case that seemed so extraordinary that the Vygotsky Circle cited it repeatedly: children's memory.

AN EXTRAORDINARY CASE: CHILDREN'S MEMORY

In their 1930 (1993) book *Studies on the History of Behavior: Ape, Primitive, and Child*, Vygotsky and Luria described memory experiments conducted by themselves and their associates. One, the "forbidden colors" experiment, had been conducted by Leontiev and would later appear in his 1933 book. In this experiment, children were asked to play a game in which they answered questions about the colors of different things. They had to avoid certain colors and use each color only once. This experiment tested their voluntary memory. First, they were tested in terms of natural (unmediated) memory. In a second round they were given mediators (cards with colors on one side) and tested again. The children tended to do better in mediated tasks, not because their natural memory improved, but because they could substitute the mediator for their own memory (Vygotsky & Luria, 1930/1993, p. 180). Yet, as the authors demonstrated (pp. 190–192), these mediators could be internalized:[2] "the process of attention still remained indirect; although instead of externally indirect, it became internally indirect" (p. 192). That is, children's *natural* memory did not improve. Instead, they developed a *cultural* memory by internalizing psychological tools that changed and systematized the task. Thus, for older children, the cards played little role in their memory. As Vygotsky summarized elsewhere, for a child, to think is to remember; for an adolescent, to remember is to think (1934/1997a, p. 99).

In terms of figure 1.1, each colored card was a mediator, a physical sign. The children quickly internalized this sign, and the internalized representation still functioned as a mediator.

Vygotsky and Luria concluded that the difference between natural (biological, evolutionary, organic) ability and cultural ability was crucial (e.g., Vygotsky and Luria, 1930/1993, p. 230; Vygotsky, 1931/1997c, p. 16; Vygotsky, 1929/1994a, p. 57). Indeed, although Vygotsky in 1925 (1974) echoed Trotsky by claiming that tool use yields both biological adaptation *and* psychological development, by the 1930s (1997a) he claimed that "the biological evolution of [humanity] was finished *before* the beginning of his historical development" (p. 97, my emphasis).

Thus, the study of memory (and other executive functions) located extraordinary abilities within the socialized human mind—specifically, in the psychological tools that individuals develop or assimilate as they learn from culturally more advanced partners. Their stance derives in part from the sixth of Marx's (1845) *Theses on Feuerbach*, which posited that the human essence emerges from "the ensemble of the social relations." This orientation toward social relations reflected the cases that Vygotsky investigated during this period: pedagogy, defectology (i.e., special needs education), rehabilitation, and an abortive fourth branch of cross-cultural psychology represented by Luria's Uzbek expedition (Luria, 1976; cf. Lamdan & Yasnitsky, 2016).

This conception of the extraordinary reflected Vygotsky's enduring interest in the Nietzschean superman reinterpreted as the New Soviet Human—an interest that was most obvious in his 1930 "The Socialist Alteration of Man" (1930/1994b) but that also appeared as early as his dissertation, *The Psychology of Art* (1925/1974), which referenced Trotsky's use of the Nietzschean superman in *Literature and Revolution* (1923/2005; see Rosenthal, 2002 for a study of how the superman trope was taken up as the New Human in Soviet culture). Read together, *Studies on the History of Behavior* chapter 1 (Vygotsky & Luria, 1930/1993), "The Socialist Alteration of [Humanity]" (Vygotsky, 1930/1994b), and "Tool and Symbol in Child Development" (Vygotsky & Luria, 1930/1994) provide a coherent argument: that psychological tools *made* humanity and could *perfect* humanity. As we will see below, Vygotsky's work was positioned as a Marxist psychology poised to reach the Soviet goal of the higher cultural and biological evolution of humanity that would be unlocked through global socialism. To solve the problem of bringing forth the New Soviet Human, Soviets would need the Vygotsky Circle's truly Marxist psychology—or so the Vygotsky Circle hoped.

To understand how Vygotsky and colleagues got to this point, let's take a step back and look at how their work began.

HISTORY

Soviet psychology was a chaotic field in the 1920s. Some psychologists—such as Georgii Chelpanov, who founded the Moscow Institute of

Psychology—believed that psychology should be separate from ideology. Others—such as Chelpanov's graduate student, Konstantin Kornilov—argued that a specifically Marxist, materialist psychology had to be developed. This disagreement came to a head at the 1923 Psychoneurological Congress, where Kornilov appealed for a Marxist psychology and Chelpanov objected in print. In the fallout, Chelpanov was dismissed from the directorship and replaced by Kornilov (Joravsky, 1989, p. 224; cf. Krementsov, 1996; Petrovsky, 1990).

Kornilov pursued a "materialist," "Marxist" line of psychology called reactology, which focused strictly on reactions and avoided issues of consciousness. Nevertheless, Kornilov allowed differing perspectives under his directorship (Van der Veer, 2014). When Vygotsky argued at the Second Neuropsychological Congress that psychological reductionism was not required by Marxism (Joravsky, 1989, p. 259), Kornilov was persuaded to hire him.

Before the 1917 Soviet Revolution, Vygotsky had faced sharply limited prospects due to his Jewish heritage. Despite garnering excellent scores, he was able to attend university only due to his luck in winning an academic lottery (Kotik-Friedgut & Friedgut, 2008; Levitin, 1982; Yasnitsky, 2018a). Vygotsky received his law degree at the University of Shanyavsky in 1917 and then returned to his hometown of Gomel to teach literature and psychology (Kotik-Friedgut & Friedgut, 2008; Yasnitsky, 2018a). At this time, he also "organised a psychology laboratory at the Gomel Teacher's College. There he delivered a course of lectures which later became a book called *Educational Psychology*" (Levitin, 1982, p. 18).

Vygotsky had reportedly always been interested in philosophical questions, assimilating Hegel by the age of fifteen (Levitin, 1982, p. 26). This education positioned him well for understanding Marx philosophically, and he became a fervent adherent of Marxism. By the time he joined the University of Moscow, Vygotsky had formulated the goal of developing a true Marxist psychology, specifically grounded in the interdisciplinary field of pedology, which focused on the development of children. This interest was not coincidental: Vygotsky's goal was to bring about the New Soviet Human (Vygotsky, 1930/1994b). He wanted to enable what Marx had termed "the all-around development of the individual" (Marx, 1875/1922) so that individuals could reach their individual and social potential in the new postcapitalist world. Vygotsky argued along with Marx that labor "contains within itself endless possibilities for the development of the human personality" (Vygotsky, 1930/1994b, p. 179) and believed that manufacturing labor plus education could yield "all-round developed people" (p. 179), "flexible" people "who would be capable of changing the forms of work, and of organizing the production process and controlling it" (p. 180).

Vygotsky quickly became the leader of a circle that included Luria and Leontiev. Both respected Vygotsky's understanding of Marxism. Leontiev later claimed that Vygotsky was the only one in the Institute who really understood Marx (quoted in Levitin, 1982, p. 117). Similarly, Luria recalled in his memoirs that Vygotsky was the leading Marxist in their troika (Luria et al., 1979, p. 41). Based on an analysis of the troika's writings in a 1925 collection, Michael Cole has argued that Luria based his Marxist approach on Marx's *Theses on Feuerbach* and Engels' *Anti-Duehring*, while Vygotsky based his directly on *Capital*—an important qualification in Soviet Russia (Luria et al., 1979, p. 204). However, some commentators have recently suggested that Vygotsky had not mastered Marx's thought to the degree that his collaborators claimed (Jones, 2018; Lamdan & Yasnitsky, 2016).

Given Vygotsky's focus on education, particularly pedology, he focused on educational contexts. And that meant that Vygotsky tended to study individual children working as subordinates under adult teachers, learning tasks at the edge of their capabilities.

Beyond consciousness, Vygotsky was intrigued by Trotsky's assertion that under socialism, "the average human type will rise to the heights of an Aristotle, a Goethe, or a Marx" (1923/2005, pp. 206–207). This focus on the New Soviet Human was popular in the Soviet Union from its early stages (Rosenthal, 2002) until the 1960s (Garcia, 2021)—a national aspiration, an accomplishment that the USSR expected to bequeath to the world.

After Trotsky was ousted from the Party in 1927, Vygotsky wisely stopped citing him. Yet the image of a New Human repeatedly made its way into Vygotsky's writings in 1930 (Vygotsky, 1930/1994b; Vygotsky & Luria, 1930/1993, 1930/1994). Vygotsky directly relates the Nietszchean superman to the New Human, reading it through his distinction between natural and cultural development:

> However, Nietzsche imagined that the development of this higher type of [human] was subject to the same law of biological evolution, the struggle for life and selection based on the survival of the fittest, which prevails in the animal world. It is for this reason that the ideal of power, the self assertion of the human personality in all the fullness of its instinctive power and ambition, rugged individualism and outstanding men and women, formed, according to Nietzsche, the road to the creation of a superman.

> This theory is erroneous, because it ignores the fact that the laws of historical evolution of [humanity] differ fundamentally from the laws of biological evolution and that the basic difference between these two processes consists of the fact that a human being evolves and develops as a historical, social being. Only a raising of all of humanity to a higher level in social life, the liberation of all of humanity, can lead to the formation of a new type of [human]. (1930/1994b, p. 182)

Drawing on Engels' account of how labor made humanity, Vygotsky goes on to argue that the New Human's biological development will follow this cultural development: "new forms of labour will create the new [human] and that this new [human] will resemble the old kind of [human], 'the old Adam', in name only" (1930/1994b, p. 183).

The cultural elevation of humanity in the pursuit of the New Human arguably led to a later expedition to Uzbekistan, which Luria conducted with Vygotsky's guidance. In this expedition, Luria hoped to find proof that literate Uzbeks were significantly altered in comparison to illiterate ones (Luria, 1976; cf. Lamdan & Yasnitsky, 2016). Even in his later holistic stage, Vygotsky remained focused on the task of raising humankind to a new plane—that is, realizing the New Human through peak or height psychology (Vygotsky, 1934/1997f; see Yasnitsky, 2018a; Zavershneva, 2016a; Zavershneva & Van der Veer, 2018, chapter 17), meant to bring human performance to the new peaks that Trotsky described.

In his development as a psychologist, as Ekaterina Zavershneva (2017) argues in her study of his archives, Vygotsky offered three distinct models of consciousness (p. 66):

- As a reflex of reflexes (1924–1926) (pp. 66–68)
- As a system of secondary connections between higher psychological functions (1927–1931) (pp. 69–74)
- As a dynamic semantic system (1932–1934) (pp. 74–78)

Zavershneva notes that Vygotsky's three models are united by three themes:

- The leading role of speech
- Systematic and semantic organization
- The origin of consciousness in the acquisition of social norms of behavior (2017, p. 92)

But as Galperin argued in 1935, this intellectual system remains incomplete: because it was speech-centric, Zavershneva summarizes, "It did not have a theory of motivation, affect, and volition" or "a well-developed theory to explain the interrelations of the personality with the environment" (Zavershneva, 2017, p. 92).

Below, we'll discuss each of these models to better situate the discussion in the historical moment. However, the remainder of the chapter will focus on Vygotsky's second model, the model that Leontiev would take up and develop as the basis of his activity theory—even after Vygotsky turned to the holistic model.

CONSCIOUSNESS AS A REFLEX OF REFLEXES (1924–1926)

As we saw in chapter 2, in his debut at the Second Psychoneurological Congress in 1924, Vygotsky acknowledged the materialist investigations of Bekhterev and Pavlov but noted that neither approach was suitable for studying consciousness. Yet, Vygotsky argued, consciousness could be studied scientifically by examining it as a *reflex of reflexes* (1997e). His argument was that "reflexes do not exist separately, do not act helter-skelter, but club together in complexes, in systems, in complex groups and formations that determine human behavior" (p. 32). Each response becomes the stimulus for a new reflex (p. 32). Thus, he argued, "consciousness is the experience of experiences" (p. 33), a "response apparatus" (p. 40) that can be conceptualized as "the reflex of reflexes" (p. 40).

Here, we see the foundation for Vygotsky's later work: the claim that lower functions (reflexes) could compound into higher, more complex ones (such as consciousness) is fundamental to all three stages of Vygotsky's work as well as Luria's neuropsychology (Luria, 1966, 1973).

On the strength of Vygotsky's 1924 presentation, Luria recommended that Kornilov take Vygotsky on at the Moscow Institute of Psychology. At the time, the Institute was attempting to develop a uniquely Marxist psychology, although its members did not agree on what such a psychology would involve. In hopes that Vygotsky would work within his reactology framework and would open up the study of consciousness, Kornilov invited Vygotsky to the Institute, where Vygotsky wrote his dissertation applying reactology to the psychology of art. When Vygotsky fell ill with tuberculosis in 1925, Kornilov waived Vygotsky's dissertation defense and enthusiastically accepted his dissertation, *The Psychology of Art* (1925/1974; cf. Toassa, 2015; Zavershneva, 2016b).

Yet despite Kornilov's enthusiasm, Vygotsky was unsatisfied with his reactological analysis. As he lay in the hospital recovering from tuberculosis,

Vygotsky drafted a manuscript called "The Cultural and Historical Crisis in Psychology," which criticized psychoanalysis, reflexology, gestalt psychology, and personalism as useful efforts that were nevertheless unable to provide a broad enough base for a materialist psychology. Vygotsky did not pursue publishing this book. Although some commentators argue that it was unpublished because it threatened the prevailing dogma of psychology (e.g., Sève, 2021), Yasnitsky draws from Zavershneva's archive work to argue that it may have been abandoned due to theoretical problems instead (2018a, pp. 56–57). In any case, Vygotsky published pieces based on this manuscript during his lifetime.

If consciousness was indeed a reflex of reflexes, what mechanism allowed human beings to control themselves? What came between the stimulus and the response? Vygotsky developed an answer to this question: *mediation*.

CONSCIOUSNESS AS A SYSTEM OF SECONDARY CONNECTIONS BETWEEN HIGHER PSYCHOLOGICAL FUNCTIONS (1927–1931)

Because of Vygotsky's focus on mediation of physical and psychological tools, this period is also known as Vygotsky's instrumentalist period (Yasnitsky, 2018a). It is during this period that Vygotsky and his circle developed many of the texts associated with Vygotsky in the West—including the text mentioned at the beginning of this chapter, the extraordinary case of children's memory. Here and elsewhere, drawing on philosophers such as Hegel (Derry, 2013) and Spinoza (Maidansky, 2018), Vygotsky postulated that the solution to the question of self-control is a third thing that comes between the stimulus and the response. In his talk at the Second Psychoneurological Congress in 1924, he had posited that each response becomes the stimulus for a new reflex. Now he pursued that line of thinking further, considering how human beings create their own responses.

To do so, and to develop a materialist, Marxist understanding of psychology that was not mechanist, Vygotsky began by criticizing behaviorism's understanding of the stimulus-response relationship. He argued that in human beings, this relationship was mediated by psychological tools, just as work is mediated by physical tools (Vygotsky, 1930/1997d; see figure 1.1). For instance, he cites Spinoza's example of Buridan's donkey, who, faced with equidistant and equally attractive alternatives, could not decide between them. Unlike the donkey, a human being in the same position might settle on a course of action by throwing a die (Vygotsky, 1997c, p. 209). The "auxiliary stimulus" of the die is a sign that allows the individual to control her or his behavior "from without" (Vygotsky & Luria, 1930/1994, p. 145).

Signs, then, are psychological tools—"artificial devices for mastering [one's] own mental processes" (Vygotsky, 1930/1997d, p. 85)—and can be considered in analogy to physical tools. But this is only an analogy, Vygotsky argued, and "cannot be carried through to the very end until all features of both concepts coincide." The point of analogical comparison is "the role these devices play in behavior, which is analogous to the role of a tool in labor": "By being included in the process of behavior, the psychological tool modifies the entire course and structure of mental functions by determining the structure of the new instrumental act, just as the technical tool modifies the process of natural adaptation by determining the form of labor operations" (Vygotsky, 1930/1997d, p. 85). Importantly, a tool can become a sign by being applied to oneself rather than to the object of one's labor (Vygotsky, 1929/1989).

This tool analogy became important for Leontiev, who, under Vygotsky's direction, conducted research on voluntary memory and attention in the 1920s, resulting in his book *The Development of Higher Forms of Memory*. This book was based on studies that were also described repeatedly in other works by the Vygotsky Circle, such as Vygotsky and Luria's (1930/1993) *Studies on the History of Behavior*.

Vygotsky and Luria wrote "Tool and Symbol in Child Development" around 1930 (see Kellogg 2011 for some detective work on the origins of this piece), and Vygotsky wrote several pieces that would become *The History of the Development of Higher Mental Functions* during 1928 to 1930 (although the book by that name was not actually published until 1960). Fragments of these works were later assembled by Luria, along with others in the same time period, and given to anglophone researchers, who combined them with illustrative texts by Vygotsky's collaborators. The result was published as *Mind in Society* (1978). Vygotsky also wrote 1930's "The Socialist Alteration of Man" and *Studies on the History of Behavior* (with Luria, also using Leontiev's memory study). They published two pieces in English at about the same time: Luria wrote "The Problem of the Cultural Behavior of the Child" in 1928, with Vygotsky publishing his companion piece "The Problem of the Cultural Development of the Child" in the same journal in 1929. Also, with Vygotsky's collaboration, Luria conducted two expeditions to Uzbekistan in which he attempted to identify differences between literate and nonliterate Uzbeks that can be traced to differences in mediation. This work was condemned at the time (Razmyslov, 1934/2000), but in 1962, almost a decade after Stalin's death, Luria pulled the manuscript out of a drawer and got it published (1976; cf. Cole's epilogue in Luria et al., 1979). All of these works emphasized the following:

- **The New Human**: The unlimited potential of humans to develop new abilities under the right conditions, specifically the conditions provided by the new Soviet regime—the "socialist alteration of [humanity]" (Vygotsky, 1930/1994b).
- **Self-mastery**: The mechanism for unlocking the unlimited potential of the New Human was humans' ability to mediate their own actions through signs (or "psychological tools"), controlling themselves from the outside. As Vygotsky said in one of his notebooks in 1928, "In Marxism, the idea of the superman . . . it is impossible to master nature without mastering oneself. Each level in the mastery of nature corresponds with a certain level in the mastery of oneself" (Zavershneva & Van der Veer, 2018, p. 117).
- **Mediation**: To master nature and self, a person had to develop and use physical and psychological tools, tools that came between the stimulus and the response (Vygotsky & Luria, 1930/1994). Physical tools mediated the person's relationship to nature, while psychological tools mediated the person's relationship to self (Vygotsky, 1930/1997d; see figure 1.3a). Vygotsky warned that although we can speak of signs as tools, it's an analogy that can't be carried through to the bitter end, and "we must not anticipate finding much similarity to working tools in these devices that we call signs" (1931/1997c, p. 60). As we saw in chapter 3, Wertsch (2007) argues that when discussing sign mediation, Vygotsky uses "mediation" in two senses: explicit, involving a visible, physical sign, and implicit, involving inner representations such as inner speech.
- **Double stimulation**: But how can a psychologist tell when an internal operation, such as voluntary attention, is mediated rather than unmediated? To detect internal mediation, Vygotsky adapted the double stimulation method. Many have attributed this method to Vygotsky, and some evidence does suggest that this method had roots in experiments that Vygotsky conducted in Gomel in 1923, before joining the Moscow Institute of Psychology. But it also had roots in the work of gestalt psychologists such as Köhler (Zavershneva & Van der Veer, 2018, p. 115; compare

Köhler, 1917/2013, which is cited extensively in Vygotsky & Luria, 1930/1993), Basov, Ach, and Janet (Valsiner, 1988, pp. 121, 134). In this method, the participant is given two stimulations with distinct functional importance, as a way to connect the "complicated internal activity" with external activity. This method is a bit like tying a string to a fish: you can't see the fish under the water, but you can see the string's movement (Vygotsky, 1994a, pp. 69–70). One example of double stimulation mentioned earlier in this chapter is that of Leontiev's forbidden colors experiment, in which children used colored cards to track which colors they could use in a game.

Throughout this stage, Vygotsky and colleagues sought to understand *higher mental functions*, voluntary executive functions mediated by culture, such as verbal thinking, logical memory, concept formation, voluntary attention, and will. These functions are based in natural *lower mental functions*. For instance, all human beings have natural memory and natural attention, but these are thoroughly changed via culture (Vygotsky 1931/1997c, pp. 17–18). The forbidden colors experiment provides one illustration, in which children used cards as cultural tools to augment their natural voluntary memory, and older children eventually internalized the functions of those cards to internally mediate their voluntary memory.

In a cultural environment, Vygotsky argued, organic development yields "a historically conditioned biological process" in which cultural development is merged with organic maturation. Thus, the human being acts upon nature—including themselves—and creates new conditions for themselves, conditions that allow the human being to shape themselves (Vygotsky 1931/1997c, p. 38). In other words, humans can use *mediation* to achieve *self-mastery*, unlocking the potential for becoming the *New Human*. Valsiner argues that "the transition of psychological functions, especially under guidance from others, into the intrapersonal sphere *in conjunction with* the development of control over these functions is the single main theme in Soviet developmental research" (1988, p. 215).

How useful is this distinction between higher and lower mental functions? Tomasello (1999) lends weight to Vygotsky's distinction, noting that "cultural processes . . . do not create new cognitive skills out of nothing, but rather they took existing individually based cognitive skills—such as those possessed by primates for dealing with space, objects, tools, quantities, categories, social relationships, communication, and social learning—and transformed them into new, culturally based cognitive skills with a social-collective dimension" (p. 7; cf. pp. 11, 51, 189). In a related vein, Toomela (2016) argues that higher psychological functions do exist, although Vygotsky does not sufficiently define them. In contrast, Arievitch (2017) cautions that "contrary to Vygotsky's distinction between the lower (unmediated) and the higher (mediated) mental functions, the totality of human development including all forms of psychological processes are culturally mediated from the very beginning of child development" (p. 57).

In any case, Vygotsky himself abandoned the distinction: "By the end of the 1920s, Vygotsky gradually came to the conclusion that 'psychological tools' cannot be 'built into' any single 'higher psychological function' because a person is an integral being and in every act of human behavior all psychological processes are manifested" (Zavershneva, 2017, p. 70). This realization led Vygotsky to attempt to explain consciousness holistically, so he introduced the principle of a system to his psychology in his paper "On Psychological Systems" (Zavershneva, 2017, p. 71; Vygotsky, 1930/1997a). From this point on, Vygotsky "repeatedly criticized his previously held views as incomplete and even erroneous" (Zavershneva, 2017, p. 74). "The idea of the sign as the mediator between nature and culture was still used as a heuristically useful abstraction, but it gradually shifted to the notion of the 'system of psychological functions'" (p. 74).

CONSCIOUSNESS AS A DYNAMIC SEMANTIC SYSTEM (1932–1934)

This systemic view led to the third model, in which Vygotsky "introduced a theoretical innovation—the notion of consciousness as a dynamic semantic system"—but could not theoretically develop it due to his untimely death in 1934 (Zavershneva, 2017, p. 74). According to Zavershneva and Van der Veer (2018), in this "principle of the semantic structure of the mind," Vygotsky retained sign mediation but "considerably broadened its concept by transitioning from the study of the external structure of the sign operation in its instrumental function to the study of its internal structure, that is, the meanings" (p. xvii). That is, Vygotsky became more interested in what Wertsch called implicit mediation (see chapter 3), and in his last book, *Thought and Language* (1934/2012, published posthumously), he focused on inner speech and the question of word meaning, which can also be translated as "meaning through language use" or "meaning through the sign operation" (Mahn, 2010, p. 315).

Zavershneva traces Vygotsky's development of this innovation from word meaning to sense to *perezhivanie* ("intellectual and emotional life experience") but notes that this "theorizing remained at the level of mere speculation, and Vygotsky's theory of consciousness was not developed any further than a sketch of a promising future theory" (2017, p. 78). The theory was heavily influenced by Lewin's field theory (Zavershneva & Van der Veer, 2018, p. xix and chapter 25). It was most fully developed in Vygotsky's notebooks, which have only recently been published (Zavershneva & Van der Veer, 2018, chapter 18).

Related to this, Vygotsky (who, with Luria, became a medical student and spent 1932–1934 studying pathology in medical settings; Van der Veer & Yasnitsky, 2016b, p. 87) became intensely interested in neuropsychology during the last months of his life, and he argued in 1934's "Psychology and the Theory of the Localization of Mental Functions" (1997b) that higher psychological functions were not localized in specific parts of the brain. As Luria (1966) later said, the brain has no reading center, no writing center. Rather, these higher functions result from networking other functions together. Vygotsky argued that any given mental function is "the product of the integral activity of strictly differentiated, hierarchically interconnected centers" (1934/1997b, p. 140). Furthermore, the brain as a whole does not connect all of these functions: "it is the product of the integral activity of dispersed, differentiated and also hierarchically interconnected functions of different brain areas" (p. 140). Thus, a local brain lesion does not damage (for instance) the brain's center of writing because such centers do not exist. Rather, it damages one of the many functional areas that are involved in the complex higher mental activity of writing (cf. Luria, 1966). Similar lesions can affect the patient differently. In development, the next higher brain structure suffers the most; in disintegration, the next lower structure does (Vygotsky 1934/1997b, p. 142; cf. Luria, 1966, pp. 36–37). Thus, Vygotsky argued that aphasia, agnosia, and apraxia represent disturbances across these connections (p. 143).

As mentioned in chapter 2, this shift from instrumental psychology to dynamic semantic systems caused considerable friction with Leontiev, who continued to pursue the instrumental approach, collapsing the difference between tools and signs (as we'll discuss in chapter 5). Zavershneva adds that "the polemic between them became increasingly pungent," a claim she substantiates by examining Vygotsky's October 1932 notebook (2016b, p. 112). "Leontiev from the early 1930s distanced himself from the study of consciousness as proposed by Vygotsky, although he continued participating in the work of the Vygotsky-Luria extended group, carried out experiments, and participated in the discussion during their symposia" (pp. 112–113). In 1932, Leontiev characterized Vygotsky as claiming that all psychology was reduced to word meaning (cited in Zavershneva, 2016b, p. 114). In turn, Vygotsky characterized Leontiev as "work[ing] near the lower boundaries" (Zavershneva & Van der Veer, 2018, p. 247).

To be fair, Leontiev had just seen Vygotsky and Luria's 1930 book criticized for being insufficiently Marxist, resulting in a critical report (Razmyslov, 1934/2000), and his own book's publication had been delayed because of these criticisms. He very well might have wanted to avoid following Vygotsky's path too closely.

So: three periods. Luria was heavily influenced by the third period and arguably continued Vygotsky's holistic approach in his neuropsychological work. Leontiev, however, stayed with the instrumental model of the second period, expanding it to the level of collective activity while simultaneously collapsing the distinction between physical and psychological tools (as we'll see in chapter 5). Since Leontiev went on to develop what we now call activity theory, in the remainder of this chapter we focus on Vygotsky's instrumental period. Below, we discuss translations in these key aspects of Vygotsky's Marxist psychology: empirical focus, mastery, and mediators.

EMPIRICAL FOCUS: INDIVIDUALS AND DYADS

When Vygotsky came to the field of psychology in 1924, the ascendant psychological research was being done by Bekhterev and Pavlov, both of whom studied reflexes. Yet, Vygotsky argued, studying these reflexes themselves did not allow us to understand consciousness—an undeniable psychological phenomenon. He proposed understanding consciousness as a reflex of reflexes, a concept that led to his mediational triangle (figure 1.1). To study mediated action, Vygotsky and other members of the Vygotsky Circle selected an empirical focus, which was an individual with a mediator, usually performing in a dyad with a more culturally advanced participant: a student with a teacher, a younger student with an older student, a preliterate individual and a literate one, or a patient needing rehabilitation and their rehabilitator.

Such dialogues are imbalanced in experience and power (cf. Tomasello, 2019, p. 166, who asserts that true dialogue emerges from peer interactions). In these unbalanced dyads, Vygotsky found the focus of his Marxist psychology. In fact, one concept most associated with Vygotsky, the zone of proximal development (see chapter 2), assumes a dyad with a more and a less culturally advanced participant.

In such dyads, agency was assumed to be an individual's. But, in keeping with Vygotsky's optimistic view of the socialist alteration of humanity, the individual was portrayed as desiring to become more culturally advanced—that is, in alignment with the more culturally advanced participant. This individual progressively learned mental functions (foremost among these, speech), and in doing so, mastered themselves more fully, building higher mental functions on lower ones.

As he moved from his instrumental period to his holistic period, Vygotsky focused more on individuals' development of consciousness, shifting from mediated action to word meaning or, more generally, sense (Vygotsky, 1927, 1934/2012; cf. Blunden, 2010; Miller, 2011). As Vygotsky argued in his last book, "Psychology, which aims at a study of complex holistic systems, must replace the method of analysis into elements with the method of analysis into units. What is the unit of verbal thought that is further unanalyzable and yet retains the properties of the whole? We believe that such a unit can be found in the internal aspect of the word, in *word meaning*" (1934/2012, p. 5).

In *Thought and Language* and elsewhere, he challenged two conceptualizations of the relationship between thinking and speaking: (1) that thought is just internalized speech and (2) that speech is just expressed thought. Rather, he argued, thought and speech do not meet their potential until they enter a dialectic relationship. At that point, they irrevocably change each other's character, though they never entirely merge. Indeed, "all the higher psychic functions are mediated processes, and signs are the basic means used to master and direct them" (p. 56). Elementary functions are subordinated during concept formation, and their "performance is mediated by word or sign" and thus interrelated (p. 114). Thought only becomes higher thought by becoming verbal, by being mediated with signs.

Vygotsky argues that the development of speech and reasoning have different roots and developmental paths. At a certain moment, these paths cross (1994a, p. 68). To explore these roots, he describes the method of double stimulation, in which the child is given two stimulations with distinct functional importance, as a way to connect the "complicated internal activity" with external activity. Thus, Vygotsky translated the empirical focus of his Marxist psychology from the vulgar material focus on reflexes to mediated action.

MASTERY: OF SELF

The focus on mediated action also accorded with a consistent theme that manifested across Vygotsky's different phases of thought: *self-mastery*, the individual's ability to exert control over themselves (Vygotsky & Luria, 1930/1993; Vygotsky, 1930/1997d, 1930/1997a, 1934/1997b). As well it should: self-mastery was an important component in Vygotsky's conception of the New Human, and one that the psychologies of Bekhterev, Pavlov, and Kornilov were ill equipped to deliver.

In fact, self-mastery underlies Vygotsky's distinction between psychological and physical tools: psychological tools (signs) allowed humans to master or control their *own behavior*, while physical tools allowed them to control or subjugate *nature* (1931/1997c, pp. 61–62). That is, psychological tools were used when the *self* was the object of one's activity, and physical tools were used when external *nature* was the object (see figure 1.3a). This overdrawn distinction shaped his instrumental period in particular, and as we've seen, became a source of disagreement with Leontiev.

One source for this claim was Engels' *Dialectics of Nature* (1971), in which he claims that labor created humanity.[3] In Engels' account, a band of apes began living on the ground rather than in the trees, thus being able to walk on hind legs and thus freeing their hands (p. 279). At that point, they could use their hands to grasp objects, which they eventually began to use as tools—and the making of tools was the first kind of labor (pp. 280–281). Labor, specifically colaboring in a collective, caused them to coevolve their minds, joint activity, and speech (which evolved as a mechanism for coordinating labor; pp. 283–284). As they cooperated, they developed higher aims, leading to the division of labor (pp. 288–289). Unlike animals, humanity began to master external labor and to premeditate, directing actions toward definite ends (pp. 290–292; cf. Marx's contrast between the bee and the architect in *Capital*).

Vygotsky and Luria took up Engels' (1971) dictum that labor created humanity, altering it to better fit their theme of self-mastery:

> 'labour created [humanity itself],' i.e. created the higher psychological functions which distinguish [humanity] as [humanity]. Primitive [humanity], using [a] stick, by means of outer sign masters the processes of [their] own behaviour and

> subordinates [their] activity to the aim which [they force] external objects to serve: tool, soil, rice. (Vygotsky & Luria, 1930/1994, p. 165; cf. Vygotsky, 1930/1994b)

That is, it was not simply the instrumental domination of nature that transformed animals into humans; it was how this focus led humans to organize higher mental functions, enabling self-mastery, mediated through signs. This self-mediatory aspect, in Vygotsky's view, had created humans and could develop them further still, so much further that they would eventually far surpass the abilities of modern humans as New Humans. For Vygotsky, the conditions for that extraordinary development could be found in socialism. In the USSR, the proletariat had risen up to rule over the bourgeoisie, as Lenin had predicted in *The State and Revolution* (Lenin, 1917/1987, drawing from Marx's 1875/1922 *Critique of the Gotha Program*), leading to a classless society. In this society, the *withering away of the state* could occur (Lenin, 1917/1987, p. 280): the functions of the government would eventually be minimized and taken over by a rotation of volunteers (Lenin, 1917/1987, p. 307). External state control would be replaced by internal mastery. And Vygotsky's Marxist psychology could help them to achieve this goal.

Echoing the concept of the withering away of the state, some Soviet educators claimed that education should be based on spontaneous processes within beneficial environments—a "withering away" of school. This notion sometimes entailed the extreme decentering of the teacher (Bauer, 1952, p. 44). Vygotsky did not endorse this view, but some Stalinist critics would associate him with it in the mid-1930s. By then, it was a dangerous view to hold: after Lenin's death, Stalin made it clear that the state would not be withering away anytime soon (Stalin, 1924/1939).

MEDIATORS: PSYCHOLOGICAL TOOLS

And this brings us to mediators themselves. As we saw above, in his instrumental period, Vygotsky introduced the mediator as a second stimulus coming between the original stimulus and the reaction.[4] An individual could consciously identify a symbolic mediator—a knot, a colored card, a mnemonic—and turn to it when encountering a stimulus. In this way, the individual could exercise self-mastery when taking a given action.

We've established that Vygotsky distinguished between psychological and physical tools. Vygotsky warned that although we can speak of signs as tools, this analogy could not be carried through to the bitter end, and "we must not anticipate finding much similarity to working tools in these devices that we call signs" (1931/1997c, p. 60; cf. 1930/1997d). What he called the "instrumental function of the sign" was "the function of stimulus-device fulfilled by the sign with respect to any psychological operation, that it is a tool of human activity" (1931/1997c, p. 60). We cannot collapse the distinction between tool and sign, he argued: "tools as devices for work, devices for mastering the processes of nature, and language as a device for social contact and communication, dissolve in the general concept of artifacts of artificial devices" (p. 61). Rather, Vygotsky argued that we should understand the difference in this way: the use of *signs* was a mediating activity in which humans regulate their own behavior (inward), while the use of *tools* was a mediating activity in which humans subjugate nature (outward) (pp. 61–62). These were "diverging lines of mediating activity" (p. 62; cf. Miller, 2011).

Similarly, in "The Instrumental Method in Psychology" (1930/1997d), Vygotsky stated that psychological tools—"artificial devices for mastering [an individual's] own mental processes"—could be considered in analogy to physical tools. But it was only an analogy, meant to highlight "the role these devices play in behavior, which is analogous to the role of a tool in labor." Such psychological tools were "artificial formations": social and "directed toward the mastery of [mental] processes" (p. 85). And "by being included in the process of behavior, the psychological tool modifies the entire course and structure of mental functions by determining the structure of the new instrumental act, just as the technical tool modifies the process of natural adaptation by determining the form of labor operations" (p. 85). That is, for Vygotsky, labor was the domain of physical tools, and behavior was the domain of psychological tools—a distinction that would be rejected in Leontiev's psychology, as we'll see in chapter 5.

Those psychological tools underpin higher mental functions, which for Vygotsky were constructed cultural abilities. Although they were *built* on natural abilities (lower mental functions), higher mental functions were thoroughly cultural, symbolic tools. For instance, logical memory was built on natural memory but exceeded it. By learning the cultural heritage of symbolic mediators, a less culturally advanced participant could build her abilities: "a human being evolves and develops as a historical, social being. Only a raising of all of humanity to a higher level in social life, the liberation of all of humanity, can lead to the formation of a new type of [human]" (Vygotsky, 1930/1994b, p. 181).

Thus, through education, human beings could acquire more cultural tools and relate them more effectively, acquiring superhuman abilities, ascending to the New Human.

SUMMARY: TRANSLATING PSYCHOLOGY TO ACHIEVE THE NEW HUMAN

As discussed, Vygotsky sought to translate existing insights into a truly Marxist psychology: not just to produce "a" Marxist psychology but to realize the vast gains in human potential that Trotsky had predicted under socialism. Vygotsky's psychology was a path to the New Human. And he carefully translated the work of his predecessors to yield this psychology.

- **Empirical focus**: He shifted from reflexes to sign mediation in individual and dyadic interactions.
- **Mastery**: He shifted from external control (inducing stimuli in subjects) to self-control (understanding how people create their own stimuli).
- **Mediation**: He shifted from physical tools (acting on the world) to psychological tools or signs (acting on oneself).

Through these translations, Vygotsky

- **problematized** pre-Marxist psychology to achieve the New Human;
- **interested** psychologists, pedologists, teachers, special educators, and others in this project;
- **enrolled** actors through various means, including academic publications, popular books (such as Vygotsky & Luria, 1930/1993), pedagogical materials, a pedology network, and both academic and popular lectures;
- **mobilized** all of these enrolled actors in the fields of psychology and pedology.

And it worked—for a short while. His vision of the New Human, whose extraordinary abilities could be unlocked by attaining psychological tools, seemed within reach. We can see Vygotsky's viewpoint on such abilities in two cases that Luria took up during and shortly after Vygotsky's death.

One case involved a truly extraordinary *ability*. Luria and Vygotsky had begun interviewing a mnemonist, who had an eidetic memory, and Vygotsky's notes suggested that he planned to write a book about the case (Zavershneva,

2016b). After Vygotsky's death, Luria continued to keep in contact with the mnemonist, eventually writing a book (Luria, 1968) describing not only how the man's extraordinary memory worked but also how it had shaped his personality. Among other things, Luria found that the mnemonist's extraordinary *natural* memory meant that he had never had to develop *cultural* tools for mediating memory that the rest of us take for granted, such as recognizing patterns that could be used for reconstructing memory.

The other case involved an extraordinary *development* of ability. Vygotsky had proposed understanding the seams between biological and cultural systems of activity by examining "deviations from the normal type," including "the so-called defective," in which natural-cultural merging does not occur normally (1931/1997c, p. 23). In the 1930s, Luria took up such a challenge when uniovular twins entered his clinic at the age of four due to their severe lack of social development. Initially, "their speech consisted only in a small number of barely differentiated sounds which they used in play and communication" (Luria & Udovich, 1956/1959, p. 39). They scarcely understood others' speech (p. 40) and did not understand symbolic play (p. 41). Applying the theories that he and Vygotsky had developed, Luria separated the twins. He identified the weaker twin (twin A)—the one who lagged developmentally—and gave that twin special activities that amounted to what we would now call speech therapy. The other twin was the "control," receiving no additional intervention. For the trained twin A, abstractions, functional categories, and logical deductions all became easier because they all depended on the mediation of speech (p. 99). When, after ten months, "both twins were set problems involving a set of operations with the aid of their own speech, it was demonstrated that an elementary special operation with the aid of speech was accessible to the trained Twin A but remained inaccessible to Twin B who had not undergone special speech training" (pp. 100–101). Neither twin would reach normal abilities, but both had improved, and twin A had improved extraordinarily, manifesting abilities that twin B could not. Twin A was hardly the New Human—but in understanding how to elevate twin A's abilities, Luria and his colleagues could discern a route to achieving the New Human.

Through his translation of psychology's empirical focus, mastery, and mediators, then, Vygotsky had constructed a Marxist psychology tuned to producing the New Human (table 4.1). He had done his best to strike a settlement that would make this psychology irresistible to other Marxist psychologists, and he had demonstrated some extraordinary results.

Table 4.1
Domain, project, translations, and trials in Vygotsky's CHAT

Domain	Psychology, pedology, defectology
Project	The New Human
Translations	
• ***Empirical focus***	Sign mediation in individual and dyadic interactions
• ***Mastery***	Exercising self-control
• ***Mediators***	Physical and psychological tools
Trials	Stalinist orthodoxy; the increasingly supervised Soviet citizen

But such settlements can always unravel—and this one did, rapidly, at the onset of the Great Terror.

HOW WELL DID VYGOTSKY'S CHAT WEATHER ITS TRIALS?

In 1931, many of Vygotsky's colleagues left for the Ukrainian Psychoneurological Academy in Kharkov, including Leontiev and Luria. Vygotsky accepted an invitation to lecture in Leningrad (Yasnitsky, 2016b, p. 39). Vygotsky and Luria also both decided to enroll in the medical department at Kharkov (Yasnitsky, 2016b, p. 32), so Vygotsky shuttled between Kharkov, Leningrad, and Moscow.

Around this time, Vygotsky and Leontiev diverged in their approach to psychology (Van der Veer & Valsiner, 1991): Vygotsky began the self-criticism that would lead to his holistic period, while Leontiev retained his orientation toward instrumentalism. Rethinking the relationship between lower and higher psychological functions, Vygotsky sought to develop "a new theory of consciousness as a dynamic semantic system" (Zavershneva, 2016b, p. 107), taking word meaning as the object of inquiry (Vygotsky, 1934/2012). Vygotsky submitted himself and his colleagues to criticism, and intensified polemic with Leontiev, who disagreed with this new turn (Zavershneva, 2016b, p. 107). Vygotsky complained that they had focused too much on the sign and not enough on the development of the operation with the sign (p. 112). Leontiev preferred to develop the instrumental line of thought that Vygotsky had abandoned, insisting that their psychology should focus on practical action (Zavershneva, 2016b, p. 112). Whereas Vygotsky had used Engels' focus on labor to point to what he considered the central phenomenon—sign mediation—Leontiev had by 1931 turned to *labor activity itself* as the central phenomenon.

So did the rest of the Kharkov Circle, which "dismissed Vygotsky's neglect of external practical activity" (Haenen, 1996, p. 26). Haenen adds that "Leontiev complained that 'it is impossible to find the cause of the development of meaning within social interaction itself' and dedicated himself to discovering 'what lies behind social interaction'" (p. 79), concluding that "the origin of consciousness had to be found in external activity" (p. 80).

As Kozulin says, the Kharkov Circle used Vygotsky's concept of internalization but disagreed with him about "the role of signs in the internalization process": "Vygotsky's emphasis on signs as means of mediation between objects of experience and mental functions was replaced by the thesis that physical action must mediate between a subject and the external world. The work of the Kharkov school established an experimental base for Leontiev's theory of the psychology of activity . . ." (Kozulin, 1984, p. 23).

We'll finish that quote in the next chapter as we trace the development of Leontiev's Marxist psychology during and after the Great Terror. For now, it's enough to say that in 1934, Vygotsky's colleagues mourned his death, published glowing obituaries in international journals, and published his last book. By 1937, they had scattered and they dared not cite his work.

Vygotsky's CHAT had failed to conform to the rapidly evolving Stalinist orthodoxy. In this orthodoxy, the Soviet citizen was expected, not to be superhuman, but to be *supervised*. In chapter 5, we'll see how Leontiev faced that trial.

5 LEONTIEV'S CHAT: MARXIST PSYCHOLOGY AFTER THE GREAT TERROR

Shortly after Vygotsky's death, the USSR descended into the Great Terror of 1936–1938. Its aftermath changed the parameters under which a Marxist psychology had to function—and thus accelerated and guided the moves Leontiev and his Kharkov Circle made away from Vygotsky's formulation. Leontiev's psychology, called *activity theory*, continued to be worked out in settings such as education and rehabilitation, but it also adhered to the necessary dictums of *partiinost'*, or party spirit, Marxism, and practicality,[1] as was required by Stalinist science (Krementsov, 1996). That is not to say that Leontiev's new settlement was Stalinist per se—but, rather, that it translated Vygotsky's instrumental psychology, with the translation more clearly serving the goals of the Stalinist USSR. To do so, it turned from Vygotsky's project of the New Human to the more modest project of a practical psychology. Further, it had to be flexible enough to respond to the many changes in the period: theorists were held up and then discarded, lines of thought rose to prominence and then were made obsolete, and CHAT had to be translated to navigate these changes. The result was due not just to Leontiev's rhetorical skill, but also to luck and happenstance: yes, Leontiev and his colleagues were savvy operators, connecting their arguments to Rubinshtein, Pavlov, and Lysenko when those figures became obligatory passage points, and discontinuing those connections when they fell out of favor. But the core of Leontiev's CHAT, tool mediation, was also fortunate enough to be ideologically unoffensive and relatively apolitical, providing a more stable settlement. In striking that settlement, Leontiev and colleagues centered their theory of activity around labor, a move that laid the foundations for CHAT's future as a framework for investigating workplaces.

To better understand Leontiev's CHAT, let's think again in terms of domain, project, translations, and trials.

- **Domain**: The domain that Leontiev addressed was psychology. Leontiev and his circle translated antecedents both from Marxist psychology and from historical materialism to help them better understand memory, perception, attention, and personality within a historical materialist framework. To do so, they borrowed the concept of mediation from Vygotsky (with a significant change that we'll discuss), the category of activity from Rubinshtein, and orthodox Marxist concepts from Engels, Lenin, Stalin, and others. These insights were translated to yield a new Marxist psychology.
- **Project**: Leontiev's project—what he offered to achieve—was not Vygotsky's superhuman but, rather, a supervised Soviet citizen, a human being who could be led into the service of the Soviet state.
- **Translations**: To produce this Marxist psychology suitable for developing supervised citizens, Leontiev translated his antecedents in several ways. We'll specifically examine translations in empirical focus, mastery, and mediators. The sum of these translations was a new settlement, activity theory.
- **Trials**: Finally, this settlement had to weather trials. In the short term, Leontiev's Marxist psychology weathered its trials far more successfully than Vygotsky's. After those trials succeeded, Leontiev later joined Luria in promoting Vygotsky's legacy both within the USSR and internationally.

In this chapter, I first provide a historical account of how Leontiev developed his psychology throughout his life, especially as he had to adapt to the changing strictures of Stalinism. Next, I examine Leontiev's psychology in terms of the three translations mentioned—empirical focus, mastery, and mediation—and how these translations led Leontiev to address his project of achieving the supervised Soviet citizen. I conclude by identifying trials that tested Leontiev's Marxist psychology—and that would eventually lead to an interventionist design approach when Leontiev's work was taken up in the West by Bødker and Engeström (chapter 6).

But first, let's start with an extraordinary case that Leontiev faced just a decade after Vygotsky's death: teaching amputees how to touch, feel, and pick up objects after losing their hands.

AN EXTRAORDINARY CASE: REHABILITATION OF AMPUTEES

In their book *Rehabilitation of Hand Function* (Leont'ev & Zaporozhets, 1945/1960), published at the end of World War II, Leontiev and Zaporozhets recounted experiments in hand rehabilitation that they conducted at a hand hospital, where they had been assigned at the beginning of the USSR's entry into World War II. In the process, they lay down markers for activity theory, which would eventually become a leading framework for Soviet psychology.

In one chapter, Leontiev reported on rehabilitation after Krukenberg's operation—an operation for someone whose hand has to be amputated. In this operation, the radius and ulna are separated and the pronator teres muscle is wrapped around both, allowing the patient to use the two bones as an elongated pincer. Obviously, this operation requires the patient to substantially reconstruct both motor and sensory impulses. In their experiments, the research team concluded that this reconstruction does not simply involve elementary sensation. Untrained patients couldn't tell if they were feeling a cube or a cylinder. But they could be trained to *interpret* these sensations (Leont'ev & Zaporozhets, 1945/1960, chapter 3).

Leontiev argued that in cases of such trauma, the motor experience is disorganized: "even when there is complete anatomical preservation of the central and peripheral system, the co-ordination of the movement may be disturbed to some degree" (Leont'ev & Zaporozhets, 1945/1960, p. 18). Thus, he argued, rehabilitation should first focus on restoring coordination (p. 18). To improve coordination, the researchers developed feedback loops for patients, reorganizing the task around different stimuli. For instance, in one experiment, they used a kymograph (an analog device that draws lines representing changes in phenomena) to provide feedback to patients as they undertook tasks with the uninjured and injured limbs (p. 19; they credit Luria's [1932/1960] work with the combined motor method). When patients had this visual feedback, they were able to smooth out their movements in mere moments (p. 21). The task had been reorganized around different stimuli: physical mediators that allowed them to reorganize their actions.

Leontiev's coauthor, Zaporozhets, noted that the motivation of activity had a large impact on outcomes: "casual and meaningless orders" could have a "chilling effect" on recovery, while "more consequential and complicated tasks" could accelerate it (Leont'ev & Zaporozhets, 1945/1960, p. 149). He even gave the example of dispirited patients reviving when they were given the meaningful task of manufacturing "window frames and furniture to replace that destroyed by the Germans at Stalingrad" (p. 150). Overall, Zaporozhets lauded "the general tonic and encouraging power of rational work activity" (p. 146), an untrammeled good providing a meaningful frame for rehabilitation. Labor, which had created humanity (Engels, 1971; Leontyev, 1959/2009b), could also rehabilitate it.

We see two important differences here between Leontiev and Zaporozhets' rehabilitation work and Vygotsky's.

- **Project**: Vygotsky had been focused on developing a Marxist psychology that could produce *superhumans*. But Leontiev & Zaporozhets instead focused on producing *supervision*. Throughout their book, rehabilitation was framed in practical, managerial terms: How could injured soldiers be rehabilitated enough to rejoin the war effort or to join the labor force? The common orientation to national motives was simply assumed: beat the Nazis, build the Soviet nation. Thus, their study located the extraordinary in common labor and cultural heritage, supervised by the state. This focus was in tune with command socialism (Lektorsky, 1999, p. 65), seeing the human being as a function for carrying out orders (Zinchenko, 1995, p. 51).
- **Translation**: Vygotsky had insisted on a clear separation between physical tools (in which a subject uses a tool to act on an object) and psychological tools (in which a subject uses a sign to act on themselves). This clear separation, like the separation between higher and mental functions that Vygotsky later renounced, was overdrawn. Using a physical tool *also* entails controlling and structuring one's own behavior, both initially (when learning to swing a hammer properly, for instance) and in an ongoing sense (when adjusting one's swing based on visual, aural, and haptic feedback). (For a more detailed discussion of this point, see Bødker & Andersen, 2005.) In the case that Leontiev and Zaporozhets described, the kymograph provided wordless feedback that the patients learned to interpret and apply to their own self-control as they practiced labor tasks (cf. McLeish, 1975, p. 176). As Leontiev argued decades later, "to handle a tool means not simply to hold it but also to know how to use it" (Leontyev, 1959/2009b, p. 192). In collapsing physical and psychological tools, Leontiev reoriented his psychology to the category of activity.

These two changes are intertwined in the history of the development of Leontiev's CHAT.

HISTORY

As with the previous chapter, this one can't really cover everything—in part because so many people were involved, and in part because the writings of many of the people involved have not been translated to English, my sole language. For instance, I acknowledge that Leontiev borrowed many ideas from Rubinshtein (Payne, 1968), but most of Rubinshtein's work is unfortunately inaccessible to me. Similarly, Gal'perin, Davydov, Zinchenko, and many others contributed to what we now know as CHAT. But let's keep our eyes on the prize: we're tracing selected major translations that lead to Engeström's CHAT.

Luria, who was internationally far more influential than Leontiev, also takes a back seat in this account. Luria's neuropsychological work, which he repeatedly credited to the ideas of Vygotsky, does not figure as heavily in Engeström's CHAT as Leontiev's does, because it did not bridge psychological and sociological analyses in the way that Leontiev's work does.

For these reasons, in this chapter we'll focus specifically on Leontiev's story and contributions—again, not as a towering figure or a genius, but as someone to whom much in CHAT has been attributed and thus whose contributions can be analyzed with a translation analysis.

As an aside, let's concede here that we do not have access to Leontiev's inner life and motivations, and we thus cannot state whether Leontiev *actually believed*, in his heart of hearts, that citizens found meaning in their lives through accepting the supervision of the Soviet state. Similarly, we do not know to what extent he agreed with Pavlov, Lysenko, or others with whom he temporarily aligned himself as those figures' fortunes rose and fell in the Soviet Union. But we are not focused on such inner motivations here: instead, we are focused on the arguments Leontiev publicly made. The supervised Soviet citizen was the project he offered, a project that could interest others in the Soviet Union, a project that made Leontiev an obligatory passage point in Soviet psychology. That story is what we cover in this chapter.

So, what is Leontiev's story? Leontiev rose to head the rehabilitation hospital through a somewhat tortuous path. This story began in the previous chapter, with the Kharkov Circle's divergence from Vygotsky—a divergence that, whatever its other merits, was politically fortuitous for Leontiev and the other Kharkovites. Below, I discuss

- how Leontiev stuck with instrumental psychology after Vygotsky abandoned it;

- how Leontiev reframed instrumental psychology to fit Stalinist doctrine;
- how Leontiev eventually reframed Vygotsky's legacy to better support his own activity theory.

INSTRUMENTAL PSYCHOLOGY: THE STONE THAT VYGOTSKY REJECTED BECAME LEONTIEV'S CHIEF CORNERSTONE

Leontiev diverged from Vygotsky in part because Leontiev wanted to keep building on the foundation of instrumental psychology—the foundation that Vygotsky had laid but later rejected in favor of a holistic psychology.

The Kharkov Circle began because in 1932, the Psychological Laboratory at the Academy of Communist Education closed in Moscow (Van der Veer & Valsiner, 1991, p. 185). Subsequently, a group of Vygotsky's students and associates (including Leontiev, Zaporozhets, and Bozhovich) established a program in developmental psychology in Kharkov, Ukraine (Kozulin, 1984, p. 22–23). They proposed a solution to the relation between consciousness and activity: consciousness develops as a child interacts with reality (Kozulin, 1986, p. xliv). By *reality*, the Kharkov Circle referred to Lenin's reflection theory, which postulated that since the psyche is material and monist, "reality is fully knowable. Truth is the reflection in the human brain of the actual objects and connections existing outside of us. . . . The fact that we know the real world is proved by human practice—that is we can transform reality by working on it to achieve a known and predictable outcome. At the same time there are no absolute, eternal, unchanging truths" (McLeish, 1975, p. 84).

This point yielded strong disagreement between the Kharkov Circle and Vygotsky (Kozulin, 1986, p. xlv) because it entailed rejecting the role of signs as chief mediators. Kozulin adds, "This is an attack not on a peripheral, but on a central notion of the cultural-historical theory [of Vygotsky]" (p. lxi). As Zinchenko argued, "'social development cannot be reduced to the history of the development of culture'" (Kozulin, 1986, p. lxi).

As we saw in chapter 4, at about this time, Vygotsky began to subject his instrumentalist work to criticism and to argue for "the principle of the semantic structure of the mind," examining signs in terms of internal meaning and semantic structure (Zavershneva & Van der Veer, 2018, p. xvii). In a December 1932 internal conference, Vygotsky proposed this new approach (p. xvii), but Leontiev was skeptical—"from a personal standpoint . . . this meeting bordered on disastrous" (Yasnitsky, 2018a, p. 117). It was a complete nonstarter for Leontiev's Kharkov Circle, which had focused on practical intelligence involving "instruments and physical objects" (p. 117). In private correspondence, Leontiev characterized Vygotsky as reducing all psychology to word meaning (Zavershneva, 2016a) and, instead, argued that his emphasis on signs should be replaced with an emphasis on labor (Van der Veer & Valsiner, 1991, pp. 289–290). In response, Vygotsky said that Leontiev's work was a step backward into the functional analysis that the Vygotsky Circle had conducted in 1928–1930, without sufficient theoretical generalization (Zavershneva, 2016a).

Gal'perin, a member of the Kharkov Circle, later claimed that "Vygotsky felt himself blocked in his cultural development" because he had trouble understanding visual representations or music: "With him, everything emerged in speech!" (Haenen, 1996, p. 28). Gal'perin believed that this state led Vygotsky to overrate speech (p. 28). In contrast, when Leontiev moved to Kharkov in 1930, he "made the concept of activity the focus of his research" (p. 78), and thus the Kharkov Circle "dismissed Vygotsky's neglect of external practical activity" (p. 26; cf. Zinchenko, 1995, pp. 39–40).

A series of events soon settled the argument:

- **1934**: Vygotsky died of tuberculosis, an event that Van der Veer sees as a blessing since otherwise his likely fate would have been the Gulag (Van der Veer, 2014, p. 29; it's unclear whether Vygotsky would have agreed). His last book, *Thought and Language*, was posthumously published by his colleagues that year. But that year also marked the beginning of the Great Purge (Fitzpatrick, 1994).
- **1935**: Leontiev delivered a lecture summarizing his critique of Vygotsky (Haenen, 1996), as discussed below.
- **1936**: The Central Committee of the Communist Party of the Soviet Union banned pedology, the interdisciplinary field with which Vygotsky was associated (see Ewing 2001 for an overview), specifically charging it with the pseudoscientific, anti-Marxist belief that children's progress depended entirely on the two factors of heredity and environment (see Leontiev, 2005, p. 58; Ewing, 2001, p. 476; for the full text of the Pedology Decree, see Wortis, 1950). The Decree called for severe criticism of pedologists and their books, and this call was immediately applied by critics to Vygotsky. Within months, Kozyrev & Turko (1936/2000) attacked Vygotsky's separation between natural and cultural development, construing it as the

heredity-environment distinction condemned in the Pedology Decree. They specifically cited *Thought and Language*.

- **1937**: Most disastrously, in January 1937, E. I. Rudneva (1937/2000) published "Vygotsky's pedological distortions," an example par excellence of Stalinist critique. Like an earlier critic, Razmyslov (1934/2000), Rudneva tied Vygotsky to positions attributed to pedology, such as mechanism, the effects of the environment, and the withering away of the school. Also like Razmyslov, she attacked Vygotsky for not grounding his work in Lenin's theory of reflection, for rooting it in word meaning rather than productive activity or labor, and for relying on bourgeois "sociologizing." And like Kozyrev & Turko, she attacked the notion that natural and cultural development could be separated.

To continue in the USSR in some form, Vygotksy's Marxist psychology had to be translated—and, necessarily, betrayed. And Leontiev had ideas about how to accomplish this.

Although Leontiev was associated with Vygotsky's circle, because he and his Kharkov Circle had already broken with Vygotsky, they were somewhat insulated from Stalinist criticisms of Vygotsky in 1936–1937. Leontiev had already collapsed the distinction between physical and psychological tools, insisting that in a Marxist psychology, the unit of analysis had to be *labor* rather than word meaning. For the Kharkov Circle, word meaning did not seem concrete or applied enough (Zavershneva, 2016a), and grounding analysis in labor was more politically viable. Whether by skill or luck, Leontiev used ideas from Vygotsky's instrumental period in ways that met the requirements of *partiinost'* (partisanship/party nature), Marxism, and practicality (cf. McLeish, 1975; Valsiner, 1988, p. 94).

Leontiev's translation has drawn accusations of betrayal. For instance, Mahn baldly states that "Leontiev developed activity theory to distance himself from Vygotsky and to make his ideals more palatable to the ruling bureaucracy" (2010, p. 316). More sympathetically, Zinchenko says that the Kharkov Circle's shift from consciousness to activity (1995, p. 39) "was obviously needed to and served to preserve, if not their lives, their freedom and, possibly, their morality" (1995, p. 40). Lektorsky allows that "the formulation and elaboration of some ideas of this theory in the 1930s and 1940s were influenced by the ideological situation in the USSR" (1999, p. 65). Translation involves understanding how interests align, which is a mutual process. Leontiev did not necessarily *choose* to translate superhumans into supervised citizens. Rather, Vygotsky's early project to produce the New Soviet Human, which once interested many, no longer interested them as much. The project of a managed, supervised Soviet citizen tended to interest many—because of Stalin's police state, because of the command economy, because World War II required people to take direction. We can read Leontiev as positioning himself as an obligatory passage point (deviate just a little, through my department of psychology, and you can achieve this), but we can also read the Soviet government as positioning itself as an obligatory passage point for Leontiev (deviate a little, apply your research interests to hand rehabilitation to support the war effort, and you can keep pursuing them).

In any case, Leontiev took a path that neatly threaded its way through the dangers of the Great Terror.

LEONTIEV TRANSLATES INSTRUMENTALIST PSYCHOLOGY INTO STALINIST TERMS

When Leontiev returned from Moscow to Kharkov's Institute for Psychology in 1937, he delivered a lecture in which he critiqued Vygotsky and reframed his own work in ways that blunted Rudneva's and others' criticisms (Keiler, 2012; Toassa, 2019). Indeed, this lecture (Leontiev, 1940/2005c) and other works Leontiev wrote around this time (Leontiev & Luria, 1937/2005; Leontiev, 1937/2005a, 1940/2005b) read as if they had used Rudneva's criticism as a checklist: the authors condemned fatalistic determination, criticized Vygotsky's use of Durkheim, rejected his focus on word meaning in favor of a focus on object-mediated labor activity, and embraced Lenin's reflection theory (reframing it in terms of internalization). Of Rudneva's list of grievances, only one major critique was not directly addressed: the false separation of natural and cultural development. This issue was indirectly addressed through Leontiev's understanding of mediation: animal activity was seen as unmediated, while human activity was mediated (Leontiev, 1940/2005a). Subsequently, Leontiev and Luria repeatedly tied their work to reflection theory (e.g., Leontiev, 1954/1957, p. 230; Luria, 1957a, p. 280) as well as to whichever Russians exemplified science at a given moment, such as Lysenko (Leontiev, 1949/1961)[2] and Pavlov.[3]

It worked. By 1939, Leontiev had recovered a supervisory position in research (Yasnitsky, 2016b, p. 46). More importantly, Leontiev's reframing of Vygotskian instrumental psychology secured its future by marrying it to the concept of activity.

LEONTIEV ADOPTS THE CONCEPT OF ACTIVITY

In shifting from Vygotsky's unit of analysis (word meaning) to labor activity, Leontiev established the foundation for what we now know as activity theory. But the concept of activity owes debts to others as well.

For instance, in her examination of Vygotsky's notebooks, Zavershneva discusses Vygotsky's later work on types of concepts, noting a possible interplay with the gestalt psychology of Kurt Lewin (cf. Lewin, 1935) via two former Lewin students who moved to Moscow and worked under Vygotsky (Zavershneva, 2017, p. 84). In a footnote, she argues that "the members of the Vygotsky Circle frequently used the conceptual apparatus of Kurt Lewin's theory and, thus, by doing this they merged Vygotskian theory with Lewin's topological and vector psychology [i.e., field theory]" (p. 85; see also Zavershneva & Van der Veer, 2018, p. xix and chapters 25–26). Yasnitsky similarly argues that activity theory was a "mutant and hybrid" of Vygotsky, Lewin, and Bernstein (Yasnitsky, 2016a, p. 225).

Another figure recognized as strongly influencing activity theory was Sergey Rubinshtein (Kaptelinin, 2015), who sat on Leontiev's 1940 dissertation committee. Rubinshtein had "transplanted" the category of activity from philosophy to psychology in 1922 (Zinchenko, 1995, p. 42). Payne (1968, chapter 5) details Rubinshtein's presentation of this idea, which clearly grounds sensation and perception in labor activity. Like the cultural-historical school and Lewin, Rubinshtein was interested in understanding psychology as inextricably social: he articulated the principle of "'unity and inseparability of consciousness and activity', according to which human conscious experience and human acting in the world, the internal and the external, are closely interconnected and mutually determine one another" (Kaptelinin, 2015). Kozulin characterizes Rubinshtein's postwar work as a "highly professional . . . presentation of Marxist philosophical anthropology" (Kozulin, 1984, p. 23).

In 1940, Leontiev both defended his dissertation (Rahmani, 1973, p. 47) and articulated the concept of activity: "a new, higher form of the reflection of reality" (Leontiev, 1940/2005b, p. 58) in which social individuals enter into relations with each other and with nature. He merged these concepts from Rubinshtein with Vygotsky's concept of mediation, characterizing activity as being mediated by tools (p. 59). Finally, he articulated the differences among activity, action, and operation (discussed below, under "Mastery: Within the Soviet State").

He would further develop these ideas when, in 1943, he was appointed as the administrator of a rehabilitation hospital specializing in hand and arm injuries, a vital function in the middle of World War II. (Also during World War II, Luria was appointed to administer a rehabilitation hospital specializing in brain injuries.) At the hand hospital, Leontiev studied how body trauma reorganizes body functions and perceptions and how articulating social motives could help injured people to reorganize these functions and perceptions more effectively in line with state goals.

After World War II, Stalinist science became more closed and chauvinistic due to the Cold War. Lysenko's August 1948 speech cemented his role as the model for Soviet science (Krementsov, 1996), and psychologists such as Leontiev and Rubinshtein were suddenly subjected to Lysenkoist criticisms (e.g., Maslina, 1950). This criticism led to Rubinshtein's ouster as chair at Moscow University in 1949, replaced by Leontiev, who had weathered the criticism better—possibly because he was ethnically Russian and thus not considered "cosmopolitan," i.e., *Semitic*, as Rubinshtein was (Kozulin, 1984, pp. 24–25). Leontiev even praised the same Pedology Decree that had launched the wave of criticism against the Vygotsky school and endangered his career (Leontiev, 1949/1961).

LEONTIEV AND LURIA REVIVE VYGOTSKY'S LEGACY—SELECTIVELY

After rehabilitating his own career by distancing himself from Vygotsky and then rehabilitating World War II veterans, in the late 1950s Leontiev turned to the task of rehabilitating Vygotsky. Like the other rehabilitations, this one supported his own position in Soviet psychology. Alexandre Metraux reports that in 1954, Leontiev and Luria were not mentioning the name of Vygotsky to their international colleagues, but by 1959, they had begun heavily promoting Vygotsky, suggesting that his works be published in French (Metraux, 2021, p. 170). From this point forward, they—and especially Luria—steadily promoted Vygotsky in the West. Metraux hypothesizes that doing this bettered their position at home compared to Pavlov (p. 173).

What happened between 1954 and 1959? In a word, de-Stalinization. After Stalin's death in 1953, several psychologists published their papers and books, including Luria and Udovich's twin study, *Speech and the Development of Mental Processes in the Child* (1956/1959), and Leontiev's *Problems of the Development of the Mind* (Leont'ev, 1959/1981), which would win the Lenin Prize in 1963. (Both cited Vygotsky.) Significantly, in 1956, Vygotsky's family also published a collection of his works. Leontiev wrote the foreword, in which he claimed that "the emphasis on semiotic mediation was transitory for Vygotsky and that the activity theory furthered the development of what was authentic in the cultural-historical school" (Kozulin, 1999, p. 253)—a claim that was only disputed after Leontiev's death (p. 254). In 1960, some of Vygotsky's writings and notes from 1928–1930, his instrumental period (Zavershneva & Van der Veer, 2018, p. xvi), were assembled to produce *The History of the Development of Higher Mental Functions* (see Yasnitsky, 2018a). A chapter-long version of the book's argument appeared in English in a 1966 collection edited by Leontiev, Luria, and Smirnov (Vygotsky, 1966). This editor-assembled book later became one of the main sources for a similar editor-assembled book, the Vygotsky book that has been most broadly cited in the West: 1978's *Mind in Society*. Closely following was Wertsch's 1981 collection, *The Concept of Activity in Social Psychology*, which included three chapters from *The History of the Development of Higher Mental Functions* that had not made it into *Mind in Society*.

In promoting Vygotsky's works, Leontiev became (for a time) Vygotsky's official interpreter, allowing him and Luria to resuscitate Vygotsky as a vital predecessor of their own work (e.g., Luria, 1966). In 1932, Leontiev had engaged in polemics with Vygotsky over Leontiev's refusal to move on from the insights of Vygotsky's instrumentalist period. Thirty years later, Leontiev de-emphasized this disagreement and emphasized the Vygotskian texts that supported his own work. Nevertheless, in shifting his focus to labor activity, Leontiev preserved Vygotsky's legacy in a highly altered form.

In 1966, Leontiev became the first dean of the Faculty of Psychology at Moscow State University. He published a final book, *Activity, Consciousness and Personality* (Leont'ev, 1975/1978), in 1975 before passing away four years later—a towering figure whose activity theory substantially influenced Soviet psychology from the late 1950s to the late 1970s, that is, during Leontiev's occupation of key administrative positions in Soviet psychology.

The stance of Leontiev's activity theory fit closely with the ideology of the bureaucratic state: a state that took a managerial viewpoint, promoting a single correct view and a narrow state-centered range of acceptable motivations for its citizens. As Blunden argues, Leontiev's functional view of society was that of a bureaucrat, a point of view that he mistook for a God's-eye view but was really from a specific social layer (Blunden, 2010, p. 209). Based on this managerial viewpoint, Leontiev in effect located the extraordinary within supervision.

As in the previous chapter, we'll discuss these key aspects of Leontiev's applied psychology: empirical focus, mastery, and mediators.

EMPIRICAL FOCUS: INDIVIDUALS WITH TOOLS

Vygotsky had grounded his units of analysis (mediated action, word meaning) in his interest in psychological as opposed to physical tools. Leontiev, however, understood meaning as emerging from labor, and thus his unit of analysis became labor activity: How did a subject use a mediator to transform the object of their labor?

This shift began in the early 1930s. Vygotsky had drawn a clear distinction between two sorts of tools:

- *Psychological* tools allowed internal mediation or self-mediation.
- *Physical* tools allowed external mediation or mediation of nature.

He saw these tools as "*diverging* lines of mediated activity" (Vygotsky, 1931/1997c, p. 62) and focused on self-mediation as the central phenomenon, making it clear that "psychological tools" was just a metaphor. But by 1931, Leontiev had turned to labor activity itself as the central phenomenon, and in so doing, had collapsed the two kinds of tools into the general category of mediation. So did the rest of the Kharkov Circle, which "dismissed Vygotsky's neglect of external practical activity" (Haenen, 1996, p. 26). Haenen adds that "Leontiev complained that 'it is impossible to find the cause of the development of meaning within social interaction itself' and dedicated himself to discovering 'what lies behind social interaction'" (p. 79), concluding that "the origin of consciousness had to be found in external activity" (p. 80). That is, while Vygotsky grounded his psychology in the sixth of Marx's *Theses on Feuerbach*, Leontiev grounded his in the first (cf. Wertsch, 1985, p. 199):

> The chief defect of all hitherto existing materialism—that of Feuerbach included—is that the thing, reality, sensuousness, is conceived only in the form of the object or of contemplation, but not as sensuous human activity, practice, not subjectively. Hence, in contradistinction to materialism, the active side was developed abstractly by idealism—which, of course, does not know real, sensuous activity as such. (Marx, 1845)

While this shift to labor activity caused a split with Vygotsky, it positioned Leontiev well for the years after the 1936 Pedology Decree. As Rudneva and others criticized Vygotsky for his supposed deviations from Marxism-Leninism, Leontiev could argue that his work was grounded in Engelsian dialectics and Lenin's reflection theory. That is, Leontiev's focus on labor activity offered a *rhetorical* solution to his Stalinist critics, one that allowed the Kharkov Circle to successfully continue under Stalinism. Yet it introduced a *methodological* problem, since it made activity both the phenomenon under investigation and the explanatory principle—a problem about which Vygotsky had warned in his unpublished 1927 methodological manuscript (Kozulin, 1984, 1999). Activity became both the map and the territory.

Leontiev expanded on the concept of activity, perhaps most influentially in his 1959 book *Problems of the Development of Mind*. In chapter 2, he—like Vygotsky and Luria and Rubinshtein before him—took up Engels' claim that labor, speech, and joint activity coevolved, that is, that humans made themselves through labor (Engels, 1971). Leontiev expanded this claim greatly, arguing four points:

- **Labor emerges in conditions of joint life.** "It was only possible for labour to emerge in animals that lived in whole groups and which had sufficiently developed forms of joint life" (Leontyev, 1959/2009b, p. 184).
- **Labor is from the beginning mediated by (1) tools and (2) social relations**, that is, in joint, collective activity: "Only through a relation with other people does [humanity] relate to nature itself, which means that labour appears from the very beginning as a process mediated by tools (in the broad sense) and at the same time mediated socially" (p. 185).
- **Speech/language originated in colaboring communication** then became separated from the immediate labor. Leontiev characterized speech as emerging from actions (labor) to yield communication (p. 195). Thus, "the production of language, and of consciousness and thought, was originally directly interwoven in production activity, in men's material intercourse" (p. 196; compare to Rudneva 1937/2000, p. 81).
- **Labor led to physical evolution.** "The origin and development of labour, this first and basic condition for the existence of [humanity], led to a change in [their] brain, [their] organs of external activity, and [their] sense organs, and their humanisation" (Leontyev, 1959/2009b, p. 181). And "[humanity]'s sense organs were also perfected through labour and

in connection with the development of the brain" (p. 183). Physical evolution reciprocally affected labor and speech: "The development of the brain and sense organs in turn had a reverse effect on labour and speech, giving both 'an ever-renewed impulse to further development'" (p. 183). "The origin and development of labour thus led to a change in [humanity's] whole physical appearance and in [their] whole anatomical and physiological organisation" (p. 184).

In Leontiev's understanding, human evolution through labor prepared humans to do two things that animals cannot:

- **Use tools collectively**: Animals do not use tools socially or collectively: their tools do not govern a community relationship or mediate productive activity. "These external means—animals' 'tools'—differ qualitatively, however, from [humanity's] true tools, the implements of labour," he explains. "The difference" is in "the *activity itself* in which they are included" (Leontyev, 1959/2009b, p. 185, my emphasis). Animals use tools for immediate action, not social activity (p. 185). But in human hands, "a tool is . . . the product of social practice and of social labour experience" (p. 193).
- **Separate object and motive**: An animal's object coincides with its motive, but a human's object and motive may diverge. By separating the object from its motive, humans separate object-oriented activity from goal-directed action (Leontyev, 1959/2009b, pp. 189–190; cf. Rubinshtein's explanation in Payne 1968, pp. 90–91). It was this modification that turned activity theory into a sub-rosa sociology. As Blunden points out, Vygotsky left us with a mass of dyads; Leontiev generated a way of talking about social formations (Blunden, 2010, p. 164). Those social formations, in true Marxist form, were built around labor, and the object was a material object that was cyclically transformed through that labor.

Yet although Leontiev could talk about social formations, his empirical research (like Vygotsky's) still investigated dyads in which one participant was more culturally advanced than the other—cases such as pedagogy, defectology, and rehabilitation. And like Vygotsky, Leontiev assumed that the less culturally advanced participant generally shared (or should be made to share) the motivations of the more culturally advanced one. Yes, he acknowledged that the less culturally advanced participant may have their own motivations as well, and in fact, participants are typically polymotivated (Leontyev, 1978/2005): a student might be motivated by an examination mark rather than by gaining productive knowledge (Leontyev, 1959/2009b). But for Leontiev, this dyadic relationship occurred within a more expansive set of structured social relations bounded by labor activity. The more culturally advanced participant represented the state, and that participant's task was to lead the individual into the state's service.

We can see this orientation in the hand rehabilitation study, in which the individuals are portrayed as disorganized people who had to be reorganized so that they could serve the state in other ways. If they could not fight the Nazis directly, they could manufacture goods for the war effort, rehabilitating themselves through labor activity (Leont'ev & Zaporozhets, 1945/1960, p. 150). Other examples include training paratroopers (Leontyev, 1978/2005) and students (Leontyev, 1959/2009b, pp. 413–414). In a 1948 piece—written just after Lysenko's triumphant speech at the August VASKhNIL meeting (Krementsov, 1996, pp. 158–159)—Leontiev emphasized the state's role in shaping students to become ideal Soviet citizens:

> But life itself, the child's activity which determines in its course his mental development, is not spontaneous—it is under the influence of education and instruction. In a Socialist society, which does not develop spontaneously but is directed by [humans], education is the decisive force which forms [humanity] intellectually. It must correspond to the aims and the needs of the entire society, of the entire people or, in other words, it must fully agree with real human needs, and also with those of individual [humans]. (Leontiev, 1948/1961, pp. 36–37)

This implies more than Western-style individual psychology, and even more than Vygotsky's dyadic psychology. It implies "a historical psychology, a theory of the historical development of the mental factor at different stages of society and in representatives of different social classes, of the basic changes in human experience produced by the abolition of private property and by the planned transformation of this experience under conditions of gradual transition from socialism to communism" (1949/1961, p. 37).

Rahmani characterizes Leontiev in this period as seeing two major tasks for Soviet psychology: "to define the structure of [human] activity through an analysis of the relationships between activity as a whole, actions and operations" and "to clarify the concept of meaning" (Rahmani, 1973, p. 47). Both tasks imply not just a psychology but a *sociology*—a point to which we will return later.

MASTERY: WITHIN THE SOVIET STATE

Like Vygotsky, Leontiev emphasized mastery in his works. But whereas Vygotsky envisioned self-mastery as yielding the New Human, Leontiev characterized mastery as inseparable from the state: proper self-mastery was in the state's service.

Leontiev characterized the subject as an individual who related to, and labored to transform, a common object. This labor occurred within an ensemble of social relationships in which the Soviet state played a critical role in guiding and developing individuals through education. In fact, Leontiev portrayed the state as consciously welcoming the child to "the society of [humanity]," specifically in how it "actively makes [the child] engage in activities." But "this does not happen without a conscious setting by society of the aims of education and instruction" (Leontiev, 1949/1961, p. 38). Elsewhere, Leontiev more pithily stated that "one is not born a personality" (quoted in Levitin 1982, p. 183) and that "a child, at birth, is a *candidate for humanity*; it cannot become human in isolation; but has to learn to become a [human] in contact with other [humans]" (Leontyev, 1959/2009b, p. 118, my emphasis). That is, one's humanity was validated not by one's biological heritage but by one's cultural heritage—by the state.

Put another way, the subject was itself the object of the state, to be transformed through education. This transformation could maximize the individual's capabilities: "Investigation of the laws governing the formation of psychic properties serves a great practical aim: the fullest possible development of the capabilities of every individual. Soviet psychologists see this as one of their most important tasks" (Leontiev, 1954/1957, p. 232). Here we see echoes of Vygotsky's peak or height psychology. But the route to that peak led through state supervision, through which the subject could learn to reorganize and self-regulate. By accepting state supervision, the subject—here, the less culturally advanced partner, working in a dyad under the supervision of the educator or rehabilitator—became capable of extraordinary things. Soldiers regained use of their limbs. Amputees could learn again how to discern shapes (Leont'ev & Zaporozhets, 1945/1960). Tone-deaf people could learn pitch (Leont'ev, 1960/1969). People could even be trained to do the *impossible*, such as sensing light with their hands (Leontyev, 1959/2009b chapter 3). These extraordinary feats were accessible to citizens who accepted the appropriate supervision and aligned their motivations with the state's.

How to analyze mastery? As we saw above, Rahmani characterizes one of Leontiev's main goals as "to define the structure of [human] activity through an analysis of the relationships between activity as a whole, actions and operations" (Rahmani, 1973, p. 47). As early as 1940 (1940/2005c), Leontiev had begun describing this structure as trifold:

- *Activity*, or collective endeavor oriented toward a shared *object* to satisfy a social *motive* (Leontiev, 1940/2005c, pp. 60–61).
- *Action*, or "a process that is directed at a conscious goal" (Leontiev, 1940/2005c, p. 62), a goal that advances the joint activity in which the individual is involved.
- *Operation*, a "social method of action" (Leontiev, 1940/2005c, p. 66) to which a tool is directed (p. 62) (For this reason, tools are considered social objects [Leontyev, 1959/2009b, p. 192]). Leontiev later characterized operations as reacting to conditions (1959/2009b, p. 155), sometimes fixed as habits (1959/2009b, p. 158).

These levels interact. For instance, when someone learns how to drive an automobile with a manual transmission, they initially treat shifting gears as an *action*, consciously attempting to reach the goal of shifting. But as they gain expertise, shifting becomes an *operation*, habitual, "as if it did not exist" (Leontyev, 1977/2009a, p. 103). Such operations can be (and often are) crystallized in the machine itself, such as when automobiles' manual transmissions were replaced by automatic transmissions (p. 103).

Leontiev argued that the structure of consciousness itself was linked to the structure of activity (Leontyev, 2009b, p. 207). Indeed, it allows us to understand human mastery at the personal level as a reflection of collective labor activity: "The realm of meanings now emerges quite differently for man. Objectively it is expressed in this, that whereas, on the one hand, there is mastery on the broadest scale of the wealth of experience of human practice crystallised and reflected in this realm, on the other hand this wealth now seems to appear to people in the light of new personal senses" (p. 238).

As this account suggests, mediators continue to be important to Leontiev's account of mastery. But Leontiev's understanding of mediators differed significantly from Vygotsky's.

MEDIATORS: TOOLS AND DIVISION OF LABOR

Between the subject and the object lay the mediator. In Leontiev's formulation, he described two kinds of mediators: tools and the division of labor. As he puts it in *Problems of the Development of Mind*, labor is from the beginning mediated by (1) tools and (2) social relations, that is, in joint, collective activity: "Only through a relation with other people does [humanity] relate to nature itself, which means that labour appears from the very beginning as a process *mediated by tools* (in the broad sense) and at the same time *mediated socially*" (Leontyev, 1959/2009b, p. 185, my emphasis).

TOOLS

As we have seen, Leontiev transformed Vygotsky's formulation by considering labor activity as the direct origination point for human psychology, thus collapsing the distinction between tool and sign, external nature and self-regulatory behavior, that Vygotsky had established.

For Leontiev, physical tools *were* psychological tools: storehouses of cultural knowledge that could be unlocked within a framing activity. They represented a cultural heritage every bit as much as Vygotsky's psychological tools. Yes, he conceded, animals also use tools, but not to mediate a community relationship or productive activity. "The difference" is in "the *activity itself in which they are included*" (Leontyev, 1959/2009b, p. 185, my emphasis). Animals use tools for immediate action, not social activity (p. 185). But in human hands, "A tool is . . . the product of social practice and of social labour experience" (p. 193). Indeed, tool mediation gives its user's action "a new structure that reflects the new objective relations: the properties of the tool, the object of labor, and the purpose of labor—its product" (Leontiev, 1967/1972, p. 153).

These physical tools could embed cultural knowledge because they functioned as congealed operations, or *crystallized activity*: "Obviously, a tool is a material object in which are crystallized methods and operations, and not actions or goals." (Leontyev, 1959/2009b, p. 102). Around the same time, he argued that psychological and physical tools alike were objectifications of embedded human activity, objectifications that could be unlocked in appropriate activity (Leont'ev, 1960/1969). "Work is crystallized" in tools; in using them, human beings consolidated the gains that they had made through their cultural heritage (p. 425). The individual's relations to the world of human objects are mediated via relations with people, including in the process of exchange. Through this process, "the individual reproduces abilities which the species *Homo sapiens* acquired in its social and historical evolution": what animals acquire through biological inheritance, humans acquire through learning (p. 426).

This view obviously represented a departure from Vygotsky, who focused on the cultural uptake of psychological tools through word meaning (e.g., Vygotsky, 1934/2012). Vygotsky was largely interested in speech as the mediator of consciousness, and even when he discussed physical tools, Vygotsky emphasized the combination of the tool's impact on nature with the sign's stimulation of behavior (Vygotsky & Luria, 1930/1993, p. 138). Moreover, "Vygotsky gradually came to the conclusion that 'psychological tools' cannot be 'built into' any single 'higher psychological function' because a person is an integral being and in every act of human behavior all psychological processes are manifested" (Zavershneva, 2017, p. 70). In contrast, Leontiev saw no practical difference between operating on nature and on oneself.

DIVISION OF LABOR

Due to his focus on labor activity, Leontiev introduced a second, separate mediator: the division of labor, whose emergence was connected to the separation of activity and action. "The appearance of goal-directed processes or actions in activity came about historically as the result of the transition of [humanity] to life in society" (Leontyev, 1977/2009a, p. 99). People in activity, he said, have their own needs satisfied via collective activity (Leontyev, 1959/2009b, p. 400). The division of labor is uniquely human (pp. 176–177) and entails dividing object from motive (p. 187). He argued that the subject—and specifically consciousness—emerges from this tension between object and motive (pp. 239–241).

For instance, consider Leontiev's famous illustration of primitive hunters who developed a division of labor to support collective activity: "It now fell to the lot of some individuals, for example, to maintain the fire and to cook food on it, and of others to procure the food itself. Some of those taking part in the collective hunt fulfilled the function of pursuing game, others the function of waiting for it in ambush and attacking it" (Leontyev, 1959/2009b, p. 186). Whereas animals' activity responds directly to their biological needs, human beings develop activity "in the conditions of a collective labour process" (p. 186)—for instance, a human being may want to hunt an animal, but his part in the hunt is to beat the bushes and frighten animals toward others in the hunting party. This *action* does not coincide with the individual's *motive* (frightening animals away seems to be the opposite of catching animals) but does achieve the individual's *object* (he shares in the spoils of the hunting party) (p. 187). The action does not make sense except in terms of joint labor activity:

> The given individual's relation with the other members of the group, by virtue of which [the individual gets a] share of the bag from them, i.e. part of the product of their joint labour activity. This relationship, this connection is realised through the activity of other people, which means that it is the activity of other people that constitutes the objective basis of the specific structure of the human individual's activity, means that historically, i.e. through its genesis, the connection between the motive and the object of an action reflects objective social connections and relations rather than natural ones. (Leontyev, 1959/2009b, pp. 188–189)

Leontiev quotes Engels' root claim: "The cause underlying the humanising of [our] animal-like ancestors is the emergence of labour and the formation of human society on its basis. 'Labour,' Engels wrote, 'created [humanity's] consciousness.'" (Leontyev, 1959/2009b, p. 181). And in Leontiev's view, changes in labor could also *remake* human consciousness in fundamental ways—specifically, in the shift from capitalist to socialist production, which leads to "a transition to a new, inner structure of consciousness, to a new 'formation' of it, viz. to the consciousness of socialist [humanity]" (1959/2009b, p. 237; cf. Luria, 1976). Like Vygotsky, Leontiev saw socialism as producing a New Human. But unlike Vygotsky, he saw labor activity as the essential unit of analysis and change, and thus the division of labor as a second essential mediator between the individual and the object. Via this angle, Leontiev and the other Kharkovites attacked what they perceived as a weak point of Vygotsky's theory, "actual relations with reality" (Kozulin, 1986, p. lx).

This move was politically shrewd. Here, I can finish that quote from Kozulin that I began in chapter 4: ". . . Leontiev's theory of the psychology of activity, which was recognized in the 1960s as an *official Soviet psychological doctrine*" (Kozulin, 1984, p. 23, my emphasis).

Leontiev's book won the Lenin Prize in 1963. Activity theory prospered from the late 1950s to the late 1970s (Kozulin, 1999).

In developing his views of these two mediators—tools and division of labor—Leontiev provided an explanation that arguably moved away from the *psychological* toward the *sociological*. Whereas Vygotsky described dyads, Leontiev described social formations (Blunden, 2010, p. 164). As Leontiev's collaborator Zaporozhets remarked in 1969, "I think that at one time we confused the psychological with the nonpsychological, i.e., with some sort of logical, sociological characterization of activity, of its basic components and the structural interrelations that occur therein, etc." (Zaporozhets et al., 1969/1995, p. 12).

So why didn't they leave it to sociology? Because *the USSR had banned sociology in 1929*, considering it a pseudoscience and recommending historical materialism instead (Osipov, 2009). This vacuum could be partially filled by Vygotsky's sociologically inflected psychology, translated by Leontiev into something that could be safely taken up during the Great Terror.

SUMMARY: TRANSLATING PSYCHOLOGY TO ACHIEVE THE SUPERVISED SOVIET CITIZEN

Leontiev, then, took elements from Vygotsky—and Rubinshtein, Engels, Lewin, and many others—to yield a Marxist psychology that could continue through the Great Terror and beyond. He translated this psychology to achieve the practical within an Engelsian understanding of the social. This angle turned out to be unpredictably fruitful. Again, Leontiev (like Pasteur [Latour, 1988]) was not a lone genius or a puppet master. Rather, he was someone who made good guesses and who was also very lucky. He translated the works of his predecessors in at least three ways:

- **Empirical focus**: He shifted from sign mediation to tool-mediated labor activity.
- **Mastery**: He shifted from individual self-control to accepting the motivations and supervision of the Soviet state.
- **Mediation**: He collapsed the tool/sign split, emphasizing tools that represented crystallized labor, as well as adding the additional mediator of division of labor.

Through these translations, Leontiev

- **problematized** psychology to fit the strictures of Marxism, *partiinost'*, and practicality;
- **interested** psychologists, rehabilitated soldiers, children, university students, and the Party;

- **enrolled** actors as far afield as Rubinshtein, Pavlov, Lysenko, and others in the Marxist science apparatus;
- **mobilized** all of these enrolled actors in the fields of psychology and physiology.

This translated psychology had to weather some unexpected trends, but the settlement held well, lasting beyond the end of the USSR itself. But in this iteration, it also depended on the internal logic of the USSR, and despite some traction in the West, it initially did not travel well beyond the USSR's borders.

In Leontiev's formulation, individuals became capable of extraordinary feats, not through unlocking new capabilities (as Vygotsky would have it) but through accepting state supervision and using the cultural inheritance of the mediators (tools, division of labor) provided to them. Individuals *became human* through engaging in collective activity in which object and motive were separated and held in tension.

These themes come together in a reminiscence that Leontiev's grandson recorded in 1978 just before Leontiev's death (Leontyev, 1978/2005). Here, Leontiev discusses the question of polymotivation, illustrating the question with a study of parachute training before World War II. The seven-story parachute tower was used (1) to train parachutists and (2) "as an amusement ride to the park's visitors" (p. 84). Leontiev added that "the jumps were absolutely safe. We had only one single accident the whole time we were working there" (p. 86). He and his collaborators tested how different conditions changed the likelihood that people would jump from the tower and then concluded that individuals were responding to both *cortical commands*, which were rational and in accordance with the training, and *subcortical commands*, which were irrational and focused on self-preservation. By changing conditions, they could appeal to the cortical level while bypassing the subcortical—for instance, by obscuring the view with paper, but explaining to the parachutist exactly what would happen—and thereby reach higher levels of compliance (pp. 85–86).

Sensitive supervision, then, could create the conditions for the individual to selflessly accept the state's goals, thereby becoming capable of extraordinary feats: confidently jumping from a seven-story building, learning to regain their range of movement, and even miraculously sensing light with their hands.

Table 5.1
Domain, project, translations, and trials in Leontiev's CHAT

Domain	Psychology
Project	The supervised Soviet citizen
Translations	
• ***Empirical focus***	Tool-mediated labor activity
• ***Mastery***	Accepting state supervision
• ***Mediators***	Crystallized labor and division of labor
Trials	Stalinist orthodoxy; tension between the theoretical and empirical; the status of activity

Through his translation of psychology's empirical focus, mastery, and mediators, Leontiev had constructed a Marxist psychology tuned to producing the supervised Soviet citizen (table 5.1). This settlement was far more durable that Vygotsky's.

HOW WELL DID LEONTIEV'S CHAT WEATHER ITS TRIALS?

Although Leontiev had arguably secured the future of the Vygotskian school by translating its insights to be taken up under Stalinism, that translation meant that some aspects of the theory were more thoroughly developed, while others were discarded.

The translation of Leontiev's CHAT into a Marxist psychology addressed the trial that Vygotsky's CHAT failed, as well as an additional trial:

- **Stalinist orthodoxy**: The continuing trial was that of conforming to the dictates of Stalinist science. Leontiev's translation of Vygotsky's cultural-historical theory into activity theory meant the survival of Vygotsky's cultural-historical school. Both Leontiev's illustrations of activity and the actual investigations were carefully noncontroversial in the Stalinist USSR. For instance, Leontiev discussed *contradictions*, or dialectical tensions in activity, but he only discussed them in capitalist societies: in Soviet societies, they supposedly did not exist. Similarly, Leontiev's investigations were of culturally less advanced individuals who could be led to align their motivations with state-approved ones, such as schoolchildren, who initially were motivated by grades and later motivated by the more culturally advanced goal of learning. Oppositional motives were not presented or examined, and certainly not validated. In Stalin's USSR, perhaps

Leontiev could not safely report results that treated oppositional motives evenhandedly or that identified tensions within the realized Utopia of the Soviet system. In any case, Leontiev's CHAT was carefully apolitical and managerial, aligning itself with whatever paradigm was dominant at any given time.

- **A tension between the theoretical and empirical**: Earlier, I quoted Zaporozhets from an internal discussion in 1969, when he argued that "we confused the psychological with the nonpsychological" (Zaporozhets et al., 1969/1995, p. 12). This confusion was reflected in the disjuncture between Leontiev's *theoretical illustrations* of activity, which were grounded in communal premodern life (hunting, fishing), and his *actual empirical investigations* of individuals and dyads in controlled conditions. Leontiev characterized social relations as grounded in concrete material labor with specific cyclical objects but did not empirically verify or trouble this characterization. His empirical studies of dyads were placed within and oriented to those labor objects. His activity theory *assumed* social relations, based on the historical materialism that was *de rigeur* in Stalinist Russia, rather than *investigating* them. As Radzikhovskii recently put it, in this conception, "it is not the individual actors who produce activity, but on the contrary, it is social activity that involves individuals and urges them to act in certain ways" (2020, p. 48). Activities acquire people.
- **The status of activity**: Was the activity a phenomenon (territory) or an analytical framework (map)? Vygotsky had argued in "Cultural and Historical Crisis in Psychology" (1927) that it could not be both. Leontiev treated this issue by ignoring the difference. In the short run, this strategy paid off. But this tension, this hidden trial, lay like an unexploded bomb that others would eventually have to defuse.

Nevertheless, activity theory began circulating across boundaries and to the West. Although Leontiev's work and that of his associates circulated to some extent, Luria's neuropsychology work circulated much farther, in part due to Luria's expertise as a scientific knowledge broker (Metraux, 2021). It brought much interest from Western psychologists, including a US graduate student named Michael Cole, who studied with Luria in 1962 as an exchange student. Luria promoted Vygotsky's works to Cole, convincing him and three other editors to pull together an English-language collection of some of Vygotsky's works: *Mind in Society* (1978). Valsiner (1988) notes that *Mind in Society* was not a Vygotsky book so much as a collection of different works by Vygotsky and his colleagues, "a potpourri . . . collected together from different sources and linked in ways judged to fit the 'recipient' culture better than the original" (p. 155). (This piecemeal composition was forthrightly acknowledged in the editors' introduction to that book.) This book was tremendously influential in the West—not least to a Finnish graduate student named Yrjö Engeström.

6 WESTERNIZED CHAT IN 1987: AN INTERVENTIONIST DESIGN APPROACH

By the time Leontiev died in 1979, *Mind in Society* (1978) had set off a Vygotsky boom in the West. Other works by Vygotsky, Luria, Leontiev, and their associates were also (linguistically) translated and taken up, first in developmental psychology and then in other fields. By the late 1980s, Leontiev's work was being applied in Northern Europe to help address two specific areas: (1) expertise and (2) human-computer interaction. These areas had both been dominated by information-processing cognitive psychology, which provided powerful insights into individual cognition but proved quite limited when addressing sociocultural issues in interventionist research (e.g., Bannon & Bødker, 1991; Engeström, 1992). The work of many Westerners figures into this history, including Michael Cole, Sylvia Scribner, Vera John-Steiner, Morten Kyng, Jean Lave, Barbara Rogoff, Roger Säljö, Sylvia Scribner, and James V. Wertsch.

But of most interest to our story is the year 1987, when two individuals in Northern Europe—Yrjö Engeström and Susanne Bødker—separately applied activity theory to research projects that were oriented to improving design in workplaces (in studies of expertise and human-computer interaction, respectively). Work was, at the time, on the cusp of great changes: new digital technologies were changing the tools that people used in workplaces, and those changes exerted pressures on how work was organized (see Spinuzzi, 2015a for an overview). In their separate projects, Engeström and Bødker responded to these rapid workplace changes by taking up and translating CHAT. Their two applications both contributed to the version of CHAT most often seen in workplace studies of expertise and human-computer interaction, which I've been calling Engeström's CHAT: a version that has been modified to guide *qualitative research* to address *collective subjects* in *multiple connected work activities* in order to support *codesigned interventions*. Although this version has since been applied to other domains, such as education and public policy, it was initially designed specifically for workplace learning, using Leontiev's framing of labor activity (for retrospectives, see Bødker, 2009; Engeström & Glăveanu, 2012).

These translations effectively refocused Westernized CHAT from the individuals or the state to workplace mediators. In the process, CHAT was translated from a psychology into a collectively oriented interventionist design approach for work, one that models work organizations and the collective difficulties they face so they can be addressed. Westernized CHAT concentrated on unlocking the extraordinary possibilities of newly formulated mediators—specifically, technologies and work organization—to revolutionize collective work activity.

Engeström and Bødker soon began to pursue CHAT along different lines, but in 1987, their ideas were new enough, and compatible enough, to leave an impression on each others' work.

To better understand Westernized CHAT in 1987, we'll again examine domain, project, translations, and trials.

- **Domain**: The domain that Engeström and Bødker addressed was interventionist design for restructuring work. They translated Leontiev's activity theory by bringing it into contact with other discourses that opposed the dominant framework of information-processing cognitive psychology (IPCP).
- **Project**: Engeström and Bødker's project was not superhumans, nor supervised citizens, but *supermediators* that could be collectively redesigned to revolutionize specific workplaces. In the majority of cases, this project involved acting with technology: developing new interfaces, texts, tools, and agreements related to them in order to unlock new capabilities at work.
- **Translations**: To produce this interventionist design approach, Engeström and Bødker translated CHAT in several ways. We'll specifically examine translations in empirical focus, mastery, and mediators. The sum of these translations was a new, Westernized CHAT, although Engeström and Bødker both characterized it simply as "activity theory."

- **Trials:** Finally, this settlement also had to weather trials. One measure of how well it weathered these trials is the proliferation of triangles in sociocultural research studies since then.

In this chapter, I first provide a historical account of how Engeström and Bødker developed CHAT around 1987, examining how they took it up in the very different networks of association in Northern Europe. Next, I examine Westernized CHAT in terms of the three translations mentioned—empirical focus, mastery, and mediation—and how these translations led Engeström and Bødker to address their project of developing supermediators. I conclude by identifying trials that tested Westernized CHAT's design approach—trials that would lead to further translations as Engeström expanded the framework into an organizational sociology (chapter 8).

TWO EXTRAORDINARY CASES: A HOSPITAL AND A COMPUTER INTERFACE

Engeström's and Bødker's uptake of CHAT was theoretically and methodologically distinct from Leontiev's (see Kaptelinin & Nardi, 2006): it guided studies in which collective subjects, in multiple connected activities, codesigned interventions intended to revolutionize their work environments. In Finland, Yrjö Engeström applied CHAT to expertise at work, while in Denmark, Susanne Bødker applied it to human-computer interaction. In both applications, participants engaged in dialogue, drawing on different experiences to codesign an emergent object, one that would become a mediator in their collective work. In doing so, they had to connect and analyze multiple activities.

To understand this Western uptake of CHAT, let's consider two cases:

- Codesigning a division of labor in a hospital
- Codesigning an interface

CODESIGNING A DIVISION OF LABOR: A HOSPITAL

Cole and Engeström (1993) describe a project in which investigators worked with Finnish hospital personnel to design a new division of labor. In the old division of labor, doctors rotated patients: when a doctor became available, they would take the next patient in line. In the new one, doctors became responsible for patients in their zones, meaning that they repeatedly saw and formed relationships with specific patients. This new division of labor yielded a new understanding of the object: *patient care* (pp. 37–39). And it was successful: "The new model dramatically changed the availability and access of care . . . there is no longer a shortage of physicians willing to work in the stations" (p. 39). Cole and Engeström described this redesign process in terms of Vygotsky's zone of proximal development (ZPD; Vygotsky, 1978), but rather than applying the ZPD concept to a dyad, they applied it to an entire collective activity system, yielding an *expansive cycle* (p. 41) in which physicians collectively deliberated and iterated the activity.

Here and in his other work, Engeström (1990, 1992, 1996a) took up CHAT to understand expertise in organizations. Engeström studied adult participants, seeing them as having already developed their own sets of expertise as fully formed agents, typically in peer- or near-peer power relationships such as medical care. His goal was to understand these encounters, work with representative participants to mutually reconfigure them, and thus improve outcomes for a general population. Through these studies, some of which were supported by the International Labor Office, Engeström would develop the *developmental work research* (DWR) approach, which focuses on how members of an activity system can collectively identify long-term contradictions and then innovate mediators to address those contradictions (Engeström, 1996b). This work laid the foundation for the interventionist research process called *change laboratory* (Engeström, 1999b; see chapter 8).

CODESIGNING A TOOL: AN INTERFACE

Meanwhile, Bødker (1991) used CHAT to underpin her work in participatory design (see Bødker, 2009 for a retrospective and Spinuzzi, 2002a and 2005a for a history). In this approach, computer scientists and workers collaborated in developing new computer systems to better support the workers' labor (e.g., Bødker et al. 1987). Thus, Bødker developed Leontiev's levels of activity to characterize collective activities, goal-directed actions, and unconscious operations (Bødker, 1991). Her goal was explicitly *not* to develop individual workers to conform to a new regime. Rather, representative participants contributed their considerable expertise to the design of new software tools, tools that could fit into the participants' existing workflows, conceptual schemes, and expertise (their social relations), supporting the population and allowing them to accomplish their work more effectively.

This goal of mutually redesigned tools and workflows was motivated in part by regulations that gave unions a say in how work was to be conducted—creating a power dynamic quite different from the top-down, hierarchical relations found in the schooling and rehabilitation settings in which Vygotsky's and Leontiev's work developed. Bødker's work, and its worker orientation, was taken up broadly within human-computer interaction and computer-supported cooperative work in the 1990s (e.g., Nardi, 1996b).

Later work brought the two traditions together: Engeström's articulation of activity systems and networks of activity systems was often combined with Bødker's articulation of the three levels of activity (Bertelsen & Bødker, 2000, 2003; Decortis et al., 2000; Engeström, 1999a; Kaptelinin & Nardi, 2012; Kuutti, 1996; Nardi, 1996b).

In both cases, extraordinary feats were portrayed as being made possible through mediators: both mediators passed down to participants and mediators consciously developed by those participants. Through the notion of crystallization forwarded by Leontiev (see chapter 5), mediators were seen as material instantiations of cognitive operations, a cultural inheritance that instantiated best practices. But in this iteration of CHAT, mediators were not just evolved and inherited: they themselves became the objects of design activity, consciously and collectively designed to solve problems in an ongoing activity. And this new relationship to mediators meant that the role of human actors changed as well. Rather than mastering a defined object through defined mediators, human actors developed and crystallized expertise as they directed their activity toward an emergent object. Whereas Vygotsky and Leontiev emphasized transforming *individuals* (via self-mastery or state mastery), Westernized CHAT now emphasized transforming *mediators* through negotiation among collective subjects to produce new collective designs and interventions, consequently reforming adjacent and intersecting activities. Whereas for Vygotsky individuals transformed themselves, and for Leontiev the state transformed individuals, for Engeström and Bødker, collectives transformed activities through mediators.

How did this shift happen?

HISTORY

As Engeström (1992) notes, Westerners such as Cole, Wertsch, and Bødker became interested in CHAT as a way to tackle sociocultural issues. Most salient to this chapter's argument is how Engeström and Bødker applied

CHAT to interventionist research in expertise and human-computer interaction (HCI), respectively.

To understand why both were interested in interventionist work, consider the place that Scandinavian democracies occupied relative to the US and Soviet blocs during these final years of the Cold War. The Scandinavian countries—including Bødker's Denmark and Engeström's Finland, as well as Iceland, Norway, and Sweden—tended to see themselves as providing a unique perspective and unique ideas. The Nordic model of democracy was considered to be a third way between U.S. and Soviet approaches (Byrkjeflot, 2001, p. 19). This model has been described as *functional socialism*: instead of abandoning property rights, Scandinavian governments made property functional to the community at large. This model represented one response to the contradiction between democracy, whose impulse generally is to empower workers, and capitalism, whose impulse is to tightly control workers so as to increase profits (Byrkjeflot, 2001, p. 22). Ehn calls this response the Scandinavian challenge (Ehn, 1989, p. 4; cf. Bjerknes et al., 1987; Spinuzzi, 2002a). As he frames it, "the Scandinavian countries are often seen as states somewhere in between capitalism and socialism, or as a bit of both. Opinions differ, but there should be no doubt that in these countries the labor movement and especially the trade unions have played a major role in the struggle for democracy and quality of work" (1989, p. 4)."

Interested in supporting workplace democracy, Marxist academics in Northern Europe began to explore links with labor unions in the 1970s. At the time, Scandinavia was about 90 percent unionized, so unions had considerable power, yet they were not well equipped to proactively guide the introduction of technologies. And that was a problem, for technologies were being introduced at a ferocious rate; corporations found that they had to introduce automation to keep up with their competitors. These new technologies imposed new ways of working that rendered workers' tacit knowledge and craft skills useless. To deal with the rapid introduction of technology and its de-skilling effects, some academics developed the collective resources approach (CRA), also known as the "critical tradition" (Bansler, 1989; Bjerknes, 1992; Kraft & Bansler, 1994). It aimed to empower labor in its struggle with management, particularly in relation to new technologies (Bansler, 1989; Nygaard, 1992). This struggle was framed in terms of democracy, and CRA research was seen as politically emancipatory (Bjerknes & Bratteteig, 1995). That focus on political emancipation and on workplace democracy implied interventionist research.

Interventionist research in Europe had other roots as well, such as Robert Jungk's future workshops (Jungk & Mullert, 1988). Jungk, an Austrian Jew who fled Nazi Germany in 1933 and attempted to alert the European media to the menace, later sought a way to conduct community-level discussions rather than ceding the future to centralized leadership. In the 1960s, Jungk started to conduct future workshops in communities:

> What is a future workshop all about? Well, the man in the street has practically no say when it comes to jobs, the environment or the way the future is shaped. All of these aspects of our lives are in the hands of the politicians, the industrialists and the experts. To remedy this non-democratic state of affairs and to show that things can be done better, Robert Jungk and his co-workers, over a decade ago, developed the future workshop method. The idea behind it is that the silent majority actually has plenty to say about what our towns and neighbourhoods should be like, about jobs and industry [and] energy and all their other needs. (Jungk & Mullert, 1988, pp. 51–52)

Jungk's future workshops were directly picked up as a technique by participatory designers (Bødker et al., 1987; Ehn, 1989), but they also bear a strong resemblance to Engeström's change laboratories. In both, deliberation among representative workers leads to a durable compromise, an agreement that comes from "below" rather than originating "above" with leadership. Since both approaches encourage deliberation, they both open up a full range of questions for participants to deliberate. In both approaches, mediators allow us to intervene at various conceptual levels, from lower-level epistemic questions such as What?, Who?, When?, and How? to higher-level questions such as Why? and Where to? (Engeström, 2007; Engeström et al., 2006; cf. Bødker & Iverson, 2002; cf. the "why," "what," and "how" layers in Bødker, 1997).

In Denmark, Bødker and her colleagues specifically sought to bring workers into the design process so that new computer tools could be *built on* their workplace expertise, serving them, rather than *replacing* that expertise and solely serving management. As Bødker says in a retrospective,

> The tradition was largely a reaction to the sociotechnical tradition imported to Scandinavia from the United Kingdom. Although this tradition placed the human being as a focus point alongside the technology in the development of organizations and technology, it neither recognized the fundamental differences and discrepancies between management and workers, nor looked at changes beyond the fine-tuning of the sociotechnical system. (Bødker, 2009, p. 277)

Early CRA projects involved educating workers about technology and organizing resistance to technologies that reduced industrial democracy. But in the UTOPIA project, the academics argued that workers had to be empowered to provide practical alternatives to management-based initiatives. That is, workers (in this case, represented by the Nordic Graphic Union, a typesetter's union) had to be able to describe a computer system that could automate work while still valuing their craft skills and upholding their autonomy. The problem was that the workers had no experience in systems design, so how could they build such systems? How could three groups (management, workers' unions, and academics) work together to develop a solution that would meet everyone's needs (Ehn, 1989; Bødker et al., 1987)?

Meanwhile, in Finland, Engeström was focused on workplace learning—a focus he took up once realizing that curricular constraints made interventionist studies in school settings difficult. As he recalled in an interview,

> I belong to the generation of the radical student movements of the 1960s and 1970s. As a university student and activist, like many others, I was dissatisfied with the then-dominant positivist, decontextualized, statistically oriented approach to education and psychology. In our search for alternatives, we found the Soviet-Russian tradition of cultural-historical activity theory, from Vygotsky to Luria, Leont'ev, Davydov, and others. I attempted to use these theories, in collaboration with progressive teachers, in intervention studies in school settings. But I found this very difficult given the curricular constraints of the time. In the world of work and organizations, I encountered a more open attitude toward novel ideas and practices. In 1982, I embarked on my first activity-theoretical study of work and learning. This was done among janitorial cleaners working for a large cleaning company. This job was commonly considered one that requires no education and that everybody can do without much cognitive effort. Inspired by Sylvia Scribner, among others, I wanted to show that every kind of work requires complex thinking, problem solving, and learning. This study was the beginning of an approach we now call developmental work research. (Engeström & Glăveanu, 2012, p. 515)

Like Bødker and her colleagues in the UTOPIA project, Engeström allied himself with labor, later publishing his *Training for Change* (1994) with the International Labour Office.

To be taken up in the Scandinavian environment, CHAT had to be translated once again. For the present argument, three aspects of the translation of CHAT to interventionist design approach are key.

- **From basic psychology to applied organizational research**: The research focus shifted from basic psychological research (or as Bakhurst says, "a fundamental explanatory strategy") to applied, practical research in organizations (or as Bakhurst says, "a method for modelling activity systems with a view to facilitating not just understanding, but practice" [Bakhurst, 2009, p. 205]).

- **From individuals to groups**: Relatedly, the empirical focus shifted from individuals in society to groups and organizations (Kaptelinin & Nardi, 2006).
- **From cognition to explicit mediators**: The location of development or change shifted from the individual's capabilities to the organization's—and thus from cognition to explicit mediating artifacts (as Miller, 2011 and Toomela, 2008 complain).

These translations were necessary, not just for application, but for allowing CHAT to disintermediate the leading psychological theory in the United States, a theory that had similarly attempted to make the transition from basic research to applied, practical research: information-processing cognitive psychology (IPCP).

IPCP

IPCP developed from the experimental tradition and posits that cognitive development is best understood in terms of processing information about the environment (as opposed to reacting to stimuli). In this tradition, the brain is metaphorically understood as a computer that processes information from the environment. As the then-dominant psychological theory in the United States, and as a theory often understood via the metaphor of mind-as-computer, IPCP seemed like a natural fit for application in realms such as artificial intelligence, human-computer interaction (HCI), and expertise. Yet some researchers who applied IPCP became discontented with its performance in these realms. As John M. Carroll argued in a retrospective, IPCP had been tested in HCI via application—and found wanting:

> The psychology of the 1970s had focused on arenas of skill and experience involving little domain-specific knowledge. Such domains were particularly amenable to information processing description by small production systems. But these theories lacked the scope and richness required for application in HCI. For example, they had little capability for modeling error or error recovery, no serious notion of cognitive development or of expertise, and they modeled human goals as unanalyzed primitives drawn from a checklist. (1991, p. 5)

Consequently, Carroll said, these theories could not be applied predictively and "could only be applied to modeling fairly limited phenomena"

(1991, p. 5). Nevertheless, these efforts continued into the 1980s, as "many cognitive psychologists began entering the HCI field, who sought to apply psychological theory (and not merely the scientific method)" (p. 5). But by the mid-1980s, the limitations of IPCP in such applications had become obvious in a variety of applications, and especially in Scandinavia, where IPCP's individualist orientation (which erased labor relations) clashed with the ideological commitment to unions and workplace democracy. Here, I review just two applications—HCI and studies of expertise—and note how these limitations drove applied researchers to take up CHAT.

BØDKER'S *THROUGH THE INTERFACE*: PARTICIPATORY DESIGN

In her search for a Marxist framework to underpin her participatory design work, Susanne Bødker—then a graduate student in computer science who was working on the UTOPIA project, in which computer scientists collaborated with typographers to develop a new interface (cf. Spinuzzi 2005a, 2005b)—picked up and applied Leontiev's CHAT to the problem of designing interfaces. The resulting doctoral thesis, *Through the Interface* (1987), was published in slightly altered form as an influential book (Bødker, 1991). (Bødker and Engeström did not know about each others' work until later in 1987 [see Bødker, 2009])

In the dissertation, Bødker cited Winograd and Flores (1986) and Dreyfus and Dreyfus (1986) to support the argument that HCI's basis in IPCP was inadequate for understanding design: "Some of the promoters of this tradition have come to the conclusion that the tradition is at present at a point, where a shift is needed from a quantitative analysis approach to a qualitative design approach: *design is where the action is in the user interface*" (Bødker, 1987, p. 2, her emphasis). She argued that "design is a process of learning" (p. 45), and to better understand it, she sought a theory of human sociocognitive development. She found it in Leontiev's CHAT.

Bødker drew on three empirical cases, including her ethnomethodology-inflected work at Xerox PARC. But most interesting to us is how she theorized the cases using Leontiev's CHAT as a starting point and applied them to the participatory design work in her work on the UTOPIA project (cf. Bødker, 1989; Bødker et al., 1987, 1991). Her interventions were inventive, often low-tech, and driven to encourage participation. For instance, she discussed mockups developed with paper and colored slides, which were designed both to educate her users about computer interfaces (most had never seen one) and to encourage their full participation (they could modify the paper mockups easily). Elsewhere, she and her collaborators call this the "tools perspective," which "takes the *work process* as its origin rather than data or information flow. This means: *not* detailed analysis, description, and formalization of qualifications, *but* the development of professional education based on the skills of professionals; *not* information flow analysis and systems description *but* specification of tools" (Bødker et al., 1987, p. 261).

The tool perspective involved recognizing the skilled craft that workers had developed over the years so that new tools would fit into that craft, rather than disrupting it: "a tool is developed as an extension of the accumulated knowledge of tools and materials within a domain" (p. 261). In the UTOPIA project, mock-ups became important in bridging the experience of researchers and workers. The researchers had begun by using traditional design description methods such as scenarios and data flow. But the graphical workers had difficulty understanding these unfamiliar notations, so the researchers and workers turned to mock-ups (p. 263)—as an interventionist technique, a "language game" (e.g., Ehn & Kyng, 1991) that would allow workers and researchers to have productive conversations.

To this end, Bødker developed three aspects of CHAT to adapt it to design work—aspects that became crucial in Westernized CHAT as an interventionist design approach.

- **Levels of activity**: First, Bødker adopted Leontiev's *levels of activity*[1] to characterize collective activities, goal-directed actions, and unconscious operations. As she explained, "The individual *activity* is conducted through a series of *actions* which have purposes in relation to the goal or object of the activity. *Operations* contribute to the realizations of actions. They are triggered by the meeting with the actual material conditions" (Bødker, 1987, p. 28, my emphasis). In characterizing activities, Bødker's goal was emphatically not to develop individual *workers* to conform to a new regime. Rather, representative participants contributed their considerable expertise to the design of new *software tools*, tools that could fit into the participants' existing workflows, conceptual schemes, and expertise (that is, their social relations), supporting the population and allowing them to accomplish their work more effectively. For this purpose, goal-directed actions and unconscious operations were well-positioned to describe Bødker's close studies of how people interacted with design artifacts, and articulating these levels also allowed her to identify and track disruptions in users' operations. The levels of activity provided her with a more fine-grained language for describing how actions and operations integrated to pursue larger objectives. Levels of activity became crucial in later HCI research as well as research in related areas (e.g., Bødker & Andersen, 2005; González et al., 2009; Kaptelinin & Nardi, 2006, 2012; Kuutti, 1996; Spinuzzi, 2003b). (The levels of activity, although mentioned by Engeström [1987/2014, chapter 2; 1990, chapter 9; 2001], largely remained peripheral to his work.)

- **Crystallization**: Second, Bødker took up Leontiev's notion of *crystallization*, in which mediators were seen as material instantiations of cognitive operations, a cultural inheritance that could instantiate best practices (Leontyev, 1959/2009b, p. 102; see chapter 5, this book). But in Bødker's implementation, mediators were not just evolved and inherited: they themselves became the objects of design activity, consciously and collectively designed to solve problems in an ongoing activity. This new relationship to mediators meant that the role of human actors changed as well. Rather than mastering a defined object through defined mediators, human actors developed and crystallized expertise as they directed their activity toward an emergent object. By transforming mediators through collective deliberation, participants could produce new collective designs and interventions.
- **Networks of activity**: Third, without using Engeström's term "networks of activity," Bødker described constellations of interconnected activities in which one activity's design object could function as another activity's mediator. Consider a writer, who labors to produce her object (a script). The script is then given to a television announcer, who uses the script as a tool in a second activity (Bødker, 1991, p. 37). The concept of networks of activity was necessary for analyzing design within interconnected systems of expertise, since HCI by definition involved applying computer expertise to domains that had not yet been computerized; that is, it involved making contact between two or more heretofore separate activities. But unlike Engeström, Bødker saw these component activities as having individual subjects.

These three aspects responded to the challenge of adapting CHAT to interventionist design. In doing so, they also responded to the limitations of IPCP.

ENGESTRÖM'S *LEARNING BY EXPANDING*: EXPERTISE

Whereas Bødker applied CHAT to HCI, Engeström applied it to workplace learning in Finland, and specifically, to designing organizational innovations to leverage and improve expertise.

Engeström found IPCP to be too confining and decontextualized to serve as a basis for understanding expertise. Inspired by Sylvia Scribner's work with adult learning and by Davydov's educational research (Engeström, 2020), he undertook his first CHAT-based study in 1982, examining how janitorial work "requires complex thinking, problem solving, and learning" (Engeström & Glăveanu, 2012, p. 515). Later, he wrote *Training for Change* (1994), a guidebook published by the International Labor Office and meant for human resources development, personnel training, and other instances in which adults are being trained. Still later, Engeström and collaborators would develop an agreement-based change process for organizations called change laboratory (Engeström, 1999b; see chapter 8). In all of these cases, Engeström's focus was on providing processes to support researchers and participants in codesigning systems of collective action.

Engeström developed these ideas in many publications, but we'll look at three early ones in particular: *Learning by Expanding* (1987/2014), *Learning, Working, and Imagining* (1990), and *Interactive Expertise* (1992).

LEARNING BY EXPANDING Engeström (1987/2014) proposed Leontiev's activity theory as a way to understand learning in the context of ongoing human activity, including work activity. This publication laid the conceptual base for *developmental work research*, a term that he had coined the same year (in a paper that he would reprint as chapter 4 in his 1990 book). However, *Learning by Expanding* did not contain work cases. Instead, it presented a dense argument based on Marx and Marxist theory, on Soviet scholars of activity theory such as Leontiev and Zinchenko, on Western CHAT scholars such as Scribner and Cole, and on Westerners studying learning, such as Bateson, Peirce, Popper, and Mead. This argument was mainly illustrated with examples from literature. In this work, he offered extensions to Leontiev's activity theory: triangles, contradictions, networks of activity, and dialogics.

Triangles Engeström provided a graphical heuristic (the now-famous triangle, figure 1.2) for picturing Leontiev's activity system. This heuristic performs a double function: (1) as a conceptual device for researchers and

(2) as an interventionist artifact used to communicate with, and codesign work with, research participants (e.g., Engeström, 1999b; Engeström & Sannino, 2010). This interplay between the two functions is critical to understand, because it explains why these triangles are distinct from the many other graphic displays (such as four-folds, matrixes, and flowcharts) that Engeström used throughout the 1980s and beyond.

1. The first function, as a **conceptual device for researchers**, is evident throughout *Learning by Expanding* (1987/2014). Here, Engeström used the triangle diagram to explain the activity system as his primary unit of analysis. In that capacity, it illustrated how labor activity emerged historically (chapter 2) and described the development of literary figures such as Twain's Huckleberry Finn and Kivi's seven brothers (chapter 3). In *Learning, Working, and Imagining*, Engeström suggested the triangle as a conceptual model for analyzing activity systems (1990, pp. 265–266) and used it to analyze a case study by Fanon (pp. 266–268) as well as his own case studies involving conversation analysis in medical work (chapters 4–6). In *Interactive Expertise*, he applied it to case studies involving conversation analysis in courtroom settings (1992, chapter 2). In these cases, the triangle provided a general conceptual model (or "theoretical instrument" [1990, p. 93]) for the academic readers, not for those who are living the activity; those people either could not participate (e.g., the fictional and historical descriptions in *Learning by Expanding*) or were not depicted as participating (in *Learning, Working, and Imagining* and *Interactive Expertise*).
2. The second function, as an **interventionist artifact for participants**, provided a means to codesign work with participants—what Ehn, following Wittgenstein, called a "language game" (Ehn 1989)—similar to the prototypes and organizational games that Bødker and other participatory designers used to leverage the tacit expertise of participants. These triangles helped participants, who did not necessarily have a grounding in theory, to make sense of their lived work. This point has been overlooked by those who have critiqued the triangle heuristic as an oversimplified conceptual tool (see Sannino, 2011 for further elaboration of this point, and see Engeström, 2018, pp. 23 and 78 for other representations Engeström has developed for local stakeholders).

This dual function meant that Engeström had to keep the triangles simple enough to perform their function as interventionist artifacts while making them complex enough to provide conceptual grounding for researchers. This latent tension between these two functions arguably makes them too simple for some researchers and too abstract for some participants. Yet perhaps it also makes them simple enough for researchers to use across many domains and fields.

Contradictions Engeström integrated philosopher Evald Ilyenkov's (1982) theorization of contradictions, making them crucial to CHAT analysis. Contradictions provided an account of change that went beyond linear development, an account that became crucial for deploying CHAT as an interventionist design approach since it allowed researchers and participants to articulate conflicts across parts of work, conflicts that were often grounded in differences in work roles and expertise.

Networks of activity Engeström expanded the analysis to networks of activity, that is, two or more interacting activity systems. He coined the term "networks of activity," theorized the concept, and understood activities as having collective subjects. This innovation was necessary for understanding interactions and contradictions between activities in workplaces as well as for modeling organizations.

Dialogics Engeström supplemented the dialectical base with the dialogism of Bakhtin (1981). While dialectics was a sufficient base for understanding learning and development in hierarchical contexts such as classrooms, dialogism was necessary in contexts in which participants of equal stature had to share perspectives and deliberate on solutions—such as in codesign work. Whereas Leontiev's CHAT provided a managerial view of activity, Engeström's CHAT drew on dialogism to facilitate multiperspectival analysis and design of work.

LEARNING, WORKING, AND IMAGINING Shortly after publishing his 1987 book, Engeström (1990) published a collection of twelve studies, most of which had been published previously between 1984 and 1990. These were mainly small-scale case studies of school-age children (e.g., chapters 1–3, 10) and medical personnel and patients (chapters 4–9). Most of the workplace studies were based on conversation analysis and coauthored by Ritva Engeström and other coauthors (chapters 5–7, 9). In this collection, Engeström names his approach "developmental work research" (chapter 4, based on a paper he originally published in 1987) and cites Bødker "in press" (what would become her 1991 book based on her 1987 doctoral thesis). Here, he argued that "contexts are activity systems" (p. 78) and provided triangular depictions of those activities as well as other displays (four-fold matrices, tables, flowcharts, other diagrammatic models) to illustrate concepts. Although the studies were mainly descriptivist, they were conducted to support interventions: "The main objective of this research," he tells us, "is not just to gain an academic response. The key objective is to produce conceptual tools for the practitioners" (pp. 104–105). In these closely bounded cases, Engeström identifies multiple mediators such as medical instruments (p. 128), medical records (chapter 7), and handwritten notes (chapter 8), as well as speculating on other potential mediators such as popular books on medicine (pp. 128–129). Perhaps because the focus was on empirical cases rather than theory, Engeström cited Marx much less in this book.

But Marx's influence was still very much in evidence. Engeström applied analyses of historical changes to work organization and work forms (1990, chapter 4) to understand how organizations and expertise had developed, drawing on the language of contradictions to characterize these changes. He briefly cited Zuboff here (p. 83) and then incorporated her work into his historical account more thoroughly in 1992's *Interactive Expertise*, along with others such as Powell and Hutchins.

INTERACTIVE EXPERTISE Strikingly, references to Marx disappeared entirely in Engeström's next book (1992). Still, this book, *Interactive Expertise*, continued the line of thought from his previous two books, identifying two perspectives on expertise: the algorithmic account, which sees expertise as residing in individuals' heads, and the enculturational account, which sees expertise and thinking as embedded in social situations, practices, and cultures (pp. 3–5). (To illustrate the latter, he quoted Dreyfus & Dreyfus, 1986.) Although these schools were typically presented as rivals, Engeström asserted that they share three propositions:

- "Expertise is universal and homogeneous."
- "Expertise consists of superior and stable individual mastery of discrete tasks and skills."
- "Expertise is acquired through internalization of experience" (Engeström, 1992, p. 5).

These propositions, Engeström argued, are Cartesian. Engeström further argued that expertise is located in activity systems rather than individuals

(Engeström, 1992, p. 11). "Expertise," he claimed, "is learning what is not yet there" (p. 14). More broadly, "expertise is here seen as an interactive accomplishment, constructed in encounters and exchanges between people and their mediating artifacts" (p. iii). To understand that accomplishment, he again drew on conversation analysis and close observation of bounded cases. These cases were interpreted as illustrating a *developmental cycle* of expert activity, which

> begins with almost exclusive emphasis on internalization, on socializing and training the novices to become competent members of the activity as it is routinely carried out. Creative externalization occurs first in the form of discrete individual deviations and innovations. As the disruptions and contradictions of the activity become more demanding, internalization takes increasingly the form of critical self-reflection, and externalization, search for novel solutions, increases.
>
> Externalization reaches its peak when a new model for the activity is envisioned, designed and implemented. As the new model becomes consolidated, internalization of its inherent ways and means again becomes the dominant form of learning and development. (Engeström, 1992, p. 16)

He characterized the developmental cycle as the collective equivalent of Vygotsky's zone of proximal development (p. 16).

Engeström also continued developing the activity system as an essential unit of analysis and networks of activity as a way of conceiving how activities interact and sometimes destabilize each other (Engeström, 1992, pp. 12–13). The studies (chapters 2–4) were largely descriptive case studies with observational and interview data, analyzed via conversation analysis, triangles, and other visual models. The third study (chapter 4) describes interventionist work as well, delivering on Engeström's earlier declaration that "the key objective is to produce conceptual tools for the practitioners" (1990, pp. 104–105).

In both Bødker's and Engeström's work—and in work that has followed this CHAT vein—empirical work took on several characteristics that were not present in previous versions of CHAT. This work

- assumed that the participants had expertise;
- involved qualitative, naturalistic studies that initially began as dyadic but quickly expanded to larger groups;
- was interventionist, treating participants as agents rather than objects, seeking to learn from their expertise and co-construct new characteristics of their activity for their benefit;

- assumed interacting activities (networks of activity);
- assumed a collective subject in Engeström's formulation (Kaptelinin and Nardi, 2006).

How did these characteristics translate CHAT for the West? Below, we discuss these key aspects of CHAT in this iteration: empirical focus, mastery, and mediators.

EMPIRICAL FOCUS: SMALL-SCALE BOUNDED CASES

As we saw in chapters 4 and 5, Vygotsky, Leontiev, and their colleagues mainly investigated individuals or dyads with unequal experience and power. But as CHAT was picked up in the West, it was applied in interventionist design research. This research required dyads or larger groups in which participants were of equal stature: codesigners who could engage in deliberative dialogue. Bødker and Engeström were not educating children or rehabilitating amputees; they were contributing expertise along with workers who were recognized as having their own considerable expertise. So, although they continued to study activities, the nature of the study was reconceptualized.

In this Westernized CHAT, as in Leontiev's CHAT, the unit of analysis was the activity. But Engeström now formulated the activity in terms of an *activity system*, in which collective subjects cyclically labor to transform a common object. The activity system was conceived as dynamic and evolving: all parts of the system might change due to internal or external contradictions, and certainly the composition of the collective subject changes over time as people enter and exit the activity.

Crucially, the activity system was applied to interventionist research in work activity. This change in application was significant, since it forced a change in the empirical focus.

Vygotsky and colleagues primarily studied dyads with clear, stable social relations, interacting in closely bounded scenes: a child and a teacher, a child and an older child, a child and a therapist. The interactants were understood as social to the core, of course, but the interactions themselves were empirically quite limited to scenes of pedagogy or rehabilitation, and the *subject* was the individual learner. Later, Leontiev and colleagues continued to primarily study dyads or individuals in experimental contexts, again with clear and stable social relations and again interacting in closely bounded scenes. Leontiev adopted the concept of activity to elaborate the social context within which the subject, still an individual learner, interacted.

In Northern Europe, though, CHAT was applied to broader and more naturalistic empirical cases, such as a doctor, nurse, and patient interacting during a clinic visit (Engeström, 1990) or a judge and two lawyers interacting in a sidebar during a court case (Engeström, 1992). Because the unit of analysis was the activity system, empirical cases were larger, although still initially quite bounded in number of participants as well as in time and space (conversations in court sidebars and medical consultations). And even when observational, as in these cases, they were oriented toward intervention: for instance, in the case of medical consultations, Engeström concluded that "our current research involves teaching teams to analyze their meetings with the conceptual tools sketched above" (p. 91). Similarly, Bødker's cases involved a worker representative and computer specialist discussing a prototype or organizational game (Bødker, 1991), with data primarily involving conversations and sometimes representative artifacts such as prototypes (Bødker, 1991) and activity system diagrams (Engeström, 1990, 1992). These cases involved meetings between activities, deliberative meetings that involved multiple participants, and consequently the subject was no longer individual but collective.

Meetings between activities meant that mediators and subjects in a given activity system could be seen as circulating from other activities or transformed by those connected activities. Thus, it became necessary to theorize contradictions *between* connected activity systems, not just *within* activities (Engeström, 1992, p. 20–22).

In shifting to design research, Engeström and Bødker required descriptive modeling of activities. Interventionist design research is conducted not *on* participants but *with* participants, oriented toward the interests of those participants, facilitating their deliberation, yielding joint emergent knowledge. This represented a fundamental reorientation from Vygotsky's and Leontiev's CHAT. Whereas Vygotsky and Leontiev dealt with individual subjects in subordinated positions (children, students, patients), Engeström and Bødker dealt with collectives in peer positions. Whereas Vygotsky and Leontiev might study an individual's use of a mediator, or provide a mediator for the individual to use, Engeström and Bødker coordinated deliberations over new mediators that the collectives would design and provided other mediators that could facilitate this deliberative design work.

As discussed above, this interventionist design approach partly resulted from two factors:

- **Political orientation**: In Finland and Denmark, the interests of capital and unions were balanced via legislation (see Spinuzzi, 2005a, 2005b).
- **The loss of lifelong employment**: In the 1980s, individuals began to enter and exit organized activities more rapidly (cf. Spinuzzi, 2015a). Consequently, developing individuals has a more limited effect on the ongoing activity than designing mediators, which can become long-term features of the activity. Mediators can be redesigned and redeveloped limitlessly, crystallizing the cultural gains of workers, and the resulting mediated activity takes the center stage in development.

In Engeström's approach, interventions involved a *cycle of expansive learning* (CEL). This cycle was grounded in the zone of proximal development—a concept discussed by Vygotsky (1930/1978) in the context of dyadic learning and then interpreted through Davydov's six learning actions (Engeström, 1987/2014, chapter 3). Engeström originally described this cycle as transforming activity through five steps (1987, chapter 3), but by 2001 he described it in seven steps:

1. Questioning
2. Historical and actual-empirical analysis
3. Modeling the new solution
4. New model
5. Implementing the new model
6. Reflection on the process
7. Consolidating the new practice (Engeström, 2001, p. 152)

The last step potentially loops back to the first for a subsequent cycle.

MASTERY: OF THE CODESIGNED MEDIATORS

Since CHAT-as-interventionist-design-approach involved deliberation among equals, it required an analytical shift from individual to collective subjects. In Engeström's and Bødker's translations, participants used their existing expertise to pursue an emergent design (that is, something that did not yet exist). That is, they learned *horizontally*, understanding the work of their collaborators, rather than just *vertically*, building new abilities on the foundations of old ones (Tuomi-Gröhn et al., 2003). Given this relationship, the collective subjects had to engage in deliberative dialogue.

The collective subject required attention to other details. Although Leontiev (see chapter 5) acknowledged that an individual subject might have multiple motives, he insisted on a single evaluative perspective. In contrast, Engeström and Bødker both assumed that subjects had multiple legitimate motives and perspectives, which were explicitly examined during the design process. For them, participants were a subset of the potential users of the resulting intervention. They represented a larger population, either politically or functionally (Spinuzzi, 2005b).

To collaborate with participants, researchers introduced "language games" (Ehn, 1989) that let the researchers and participants collaboratively describe the existing system and the problems within it. Such language games can include participatory design techniques such as prototyping and organizational games as well as techniques such as constructing, examining, and discussing activity system diagrams, which serve as a "mediating second stimulus" for the participants (Engeström & Sannino, 2010, p. 15). On the basis of such common, consensual descriptions, participants and researchers can then generate and test potential futures. This point is critical for understanding Engeström's CHAT, and we'll discuss it further in the section "Mediators: Tools, Rules, and Division of Labor."

Engeström, Bødker, and others in this CHAT tradition agree that individual development occurs. But unlike in Vygotsky's or Leontiev's CHAT, the individual is not the focus of the transformation—their collective social interaction is. We'll see how this expansion plays out in chapter 8.

MEDIATORS: TOOLS, RULES, AND DIVISION OF LABOR

In this translation, mediators include

- tools (which mediate between the subject and object);
- rules (which mediate between the subject and a community);
- division of labor (which mediates between a community and its object) (see figure 1.2).

These mediators are analytically separate, but they are recognized as interacting in a bounded case. Their relationships are modeled in Engeström's triangle to identify their interactions and to provide a representation (language game) that researchers and participants alike can understand and discuss.

To redesign these mediators in a work activity, codesigners have to talk about the current, undesirable state of the work activity. They use Engeström's triangle and other language games to do so. That is, they *design* new mediators by using *designing* mediators:

- **Designed mediators**: Tools, rules, and divisions of labor that collective subjects develop to support their work, such as the new division of labor in a hospital or the new computer interface for typographers. Engeström and Bødker were both interested in redesigning existing mediators to better support work in the activity being studied; above, I characterized these redesigned mediators as *supermediators* to emphasize their special status as revolutionizing that target activity.
- **Designing mediators**: Those supermediators themselves had to be designed through a separate designing activity, an activity in which the supermediator was itself the object. In that designing activity, how could researchers and participants join to codesign this new tool or division of labor? They used novel mediators that were developed specifically for design work: representations (language games) that collective subjects used as they deliberated over their work and the work needs for which they had to design future mediators. These mediators included activity system diagrams, prototypes, and organizational games.

This second class of designing mediators was new to CHAT—they can't be found in Vygotsky or Leontiev—and they reflected Westernized CHAT's new orientation. We can see their relationship in figure 6.1, a basic activity network diagram.

What theoretical understanding underpinned both designing and designed mediators? Engeström and Bødker both drew from Leontiev's understanding of tools as crystallized operations passed down as a cultural inheritance (Leontyev, 1959/2009b, p. 102; Leont'ev, 1960/1969). But Leontiev did not explore design activity, so Engeström and Bødker had to draw on other sources, such as Ehn's design language.

Such borrowing was critical. In Scandinavia, design work was the purview not of state supervisors but, rather, of the workers who would use the resulting designs, supported by their unions. Thus, collective design meant understanding several activities and their relationships: a current activity,

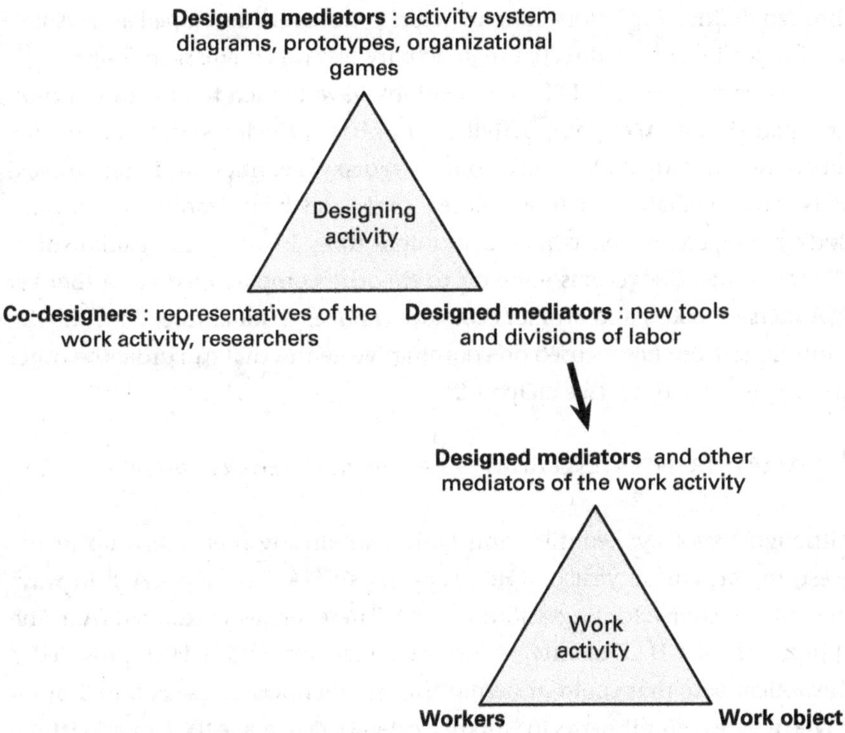

FIGURE 6.1
In the interventionist designing activity (top), codesigners use *designing mediators* to create *designed mediators* meant to be used in the work activity (bottom).

a future activity, other activities that might be affected, and finally, the design activity being undertaken to design new tools or divisions of labor. Hence, Engeström developed the concept of networks of activity and drew on Ilyenkov to discuss external contradictions as well as internal ones.

Tools were understood as crystallized labor that was intentionally and dialectically codesigned. Crystallization was especially used in participatory design (Bannon & Bødker, 1991; Bertelsen & Bødker, 2000; Bødker & Petersen, 2000; Bødker & Klokmose, 2011; Spasser, 2000), where it provided a way to conceptualize how interface elements could both frame actions and embed expertise. In Bødker's implementation, such mediators were not just inherited, but also consciously and collectively designed to solve problems in an ongoing activity. This new relationship to mediators meant that the role of human actors changed as well: rather than mastering a defined object through defined mediators, human actors collectively developed and crystallized expertise as they directed their activity toward an emergent object.

Since Westernized CHAT investigations have tended to encompass multiple participants in ongoing activities, they have broadened the set of mediators under investigation at any point. Vygotsky, Leontiev, and Luria looked at isolated mediators such as colored cards and kymographs. In contrast, Bødker and Engeström explored multiple tools in complex relationships. Consequently, Bødker has gone on to theorize complex mediation (Bødker & Andersen, 2005) and artifact ecologies (Bødker & Klokmose, 2011, 2012), while Engeström has focused on coordinative genres that can order the other artifacts in an activity (see chapter 8).

SUMMARY: TRANSLATING CHAT TO ACHIEVE SUPERMEDIATORS

Although Vygotsky, Leontiev, and Luria had already been taken up in the West, in this crucial year of 1987, Leontiev's CHAT was translated in ways that set the course for Engeström's CHAT. These key ways resulted from the application of CHAT to interventionist design research: CHAT provided a theoretical base that could underpin this interventionist research in Scandinavia, where commitments to workplace democracy made IPCP unappealing.

But in taking up CHAT, Bødker and Engeström translated it in key ways:

- It became oriented to the design of mediators in organizations rather than to the behavior of individuals in society.
- It became dialogically interventionist, with participants functioning as active deliberators rather than relatively passive study subjects.
- It treated these participants as agents rather than objects.
- It introduced language games that could put researchers and participants on the same footing, using heuristics and other designing mediators to facilitate dialogue and deliberation.

These changes can be summed up in three different translations:

- **Empirical focus**: CHAT's empirical focus shifted from tool-mediated labor activity in dyads to reflections on and deliberations about labor activity in larger groups.
- **Mastery**: Its understanding of mastery shifted from accepting state motivation to participating in the design of mediators.

- **Mediators**: Its mediators themselves expanded beyond tools and division of labor to include rules; and the focus of research activity shifted from understanding how people learn to use mediators to facilitating codesign of *designed* mediators via *designing* mediators.

Critically, looking back on chapters 4–6, *the different translations of CHAT focused on what the researchers could control.* Vygotsky could control children and uneven dyads, and his research mainly took place in schools and rehabilitation centers, where he had some say over the activities in which his participants engaged. That control was greatly diminished after the Pedology Decree, but Leontiev managed to find similar settings to control participants in his own rehabilitation facility, in schools, and in other experimental settings. But Bødker and Engeström could not control their participants—workers—who were understood as equal partners in the interventionist research. Instead, Bødker and Engeström worked with their participants to control other aspects of the work activity: tools, rules, and division of labor.

Through these translations, Bødker and Engeström separately

- **problematized** information-processing cognitive psychology and interventionist design;
- **interested** unions, management, and academic researchers in the broader Scandinavian commitment to workplace democracy;
- **enrolled** CHAT forebears (Vygotsky, Leontiev, Davydov), Western cultural psychologists, Jungk, and others focused on transformational participation;
- **mobilized** these enrolled actors to carry out interventionist design research across fields, creating research settings in which the codesigned mediators themselves were the key result of the research, and the codesigners themselves were the beneficiaries.

Through their translation of empirical focus, mastery, and mediators, Bødker and Engeström constructed an interventionist design approach tuned to codesigning supermediators (table 6.1). This settlement was very different from Vygotsky's and Leontiev's, and though it did not dominate psychology as Leontiev's activity theory did, it was also no longer confined to psychological study: it spread quickly to multiple fields, where it would be taken up and translated. It explored more expansive empirical settings: a broader context, bounded by the object of the activity under investigation. But context is infinitely expandable, as we'll see in chapter 8.

Table 6.1
Domain, project, translations, and trials in Westernized CHAT

Domain	Design
Project	Emancipation of labor
Translations	
• ***Empirical focus***	Deliberations about labor activity in larger groups
• ***Mastery***	Participating in the design of mediators
• ***Mediators***	Rules; codesign of designed mediators; introduction of designing mediators
Trials	Abbreviated scope; changes in work; stable sets of interlocutors; the status of activity

HOW WELL DID WESTERNIZED CHAT WEATHER ITS TRIALS?

The turn to interventionist design research introduced new trials:

- **Abbreviated scope relative to design aspirations**: Although activity systems and networks of activity provided a way to model organizations, the scope of investigation was still focused on small numbers of participants in tightly bounded scenes. This empirical scope was larger than that of the previous iterations, but it was too small to truly investigate or model organizations, especially in terms of historical development and continuity.
- **Changes in work**: Engeström (1987/2014) characterized activities in terms of historically emerging forms of work: craft activity, rationalized activity, humanized activity, and collectively and expansively mastered activity. He noted that work was becoming more networked (1992). But these changes were not captured well, again due to the abbreviated scope.
- **Stable sets of interlocutors**: In applying CHAT to work rather than educational or rehabilitation settings, Westerners lost the stable set of interlocutors implied in the other settings. This loss was not immediately obvious. The interventionist stance of this iteration of CHAT assumes a stable set of stakeholders who can serve as codesigners as they engage in a dialectic, juxtaposing contradictory viewpoints or ideas to develop a common settlement. Yet those stable stakeholders are not guaranteed in cases such as social and peer production. As we'll see near the end of chapter 8, much work is oriented to projects in which fluid sets of

stakeholders participate intermittently, rotate leadership roles, and may define and transform the object in different and conflicting ways.

- **The status of activity**: Was the activity a phenomenon (territory) or an analytical framework (map)? This tension, left over from Leontiev's CHAT, was not resolved in the new iteration of CHAT.

The first two of these trials were to be resolved in the work that Engeström conducted with colleagues at the Center for Research on Activity, Development, and Learning (CRADLE, established in 1994) at the University of Helsinki from 1996 on. This work transformed CHAT into a de facto organizational sociology—and into what we can finally, unambiguously call *Engeström's CHAT*.

7 INTERLUDE: CHAT AS A SOCIOCOGNITIVE BASE FOR NORTH AMERICAN WRITING STUDIES

Up to now, we have looked at major translations that led to Engeström's CHAT, and we have centered this narrative around individual figures—not as lone Machiavellian geniuses who steer history, but as people to whom significant actions have been attributed. But innovations such as CHAT do not have unitary identities. They are taken up in many instances, for many purposes, often becoming localized in how they are understood and implemented in specific locations, disciplines, fields, and subfields.

This chapter describes one such instance, one localized uptake of CHAT for specific purposes: in my own field, writing studies. I call this chapter an *interlude* since this heterodox localized uptake has not significantly impacted Engeströmian CHAT beyond this field. Yet that is true of many fields in the decade after 1987, especially those in which researchers and practitioners sought new ways to conceptualize and explore the consequences of rapid technological change at work: human-computer interaction, computer-supported cooperative work, user experience design, information systems, and professional communication. For such fields, Engeström's CHAT came along at the right time, with the right features, to address a developing need.

Writing studies also happens to be *my* field, and the changes described here happened when I was in graduate school and shortly afterward. So this chapter, unlike the others, is somewhat autobiographical. It draws heavily on my own experiences in workplace writing studies, and it foregrounds the needs in my specific subfield, professional communication. In that subfield, we were undergoing rapid growth due to the PC revolution, in which people across multiple industries had to learn how to work with computers,

computers whose interfaces were not only crude but also often wildly different from application to application. This meant that software publishers needed to generate specialized documentation for these applications, and this documentation could not simply describe features; it had to anticipate the needs of situated users who were attempting to perform specific tasks. The need for this documentation drove extraordinary growth in US professional communication programs in the late 1980s and throughout the 1990s. When I began my Ph.D. studies in rhetoric and professional communication in 1994, that program was one of only a dozen offering a Ph.D. in that field in the United States. By 2003, that number had doubled. Growth was far more rapid at the MA and BA levels.

Yet in professional communication, and in writing studies more generally, empirical research lagged. Such programs were largely situated in English departments, and their practitioners were largely trained in literary analysis rather than in empirical research traditions. In the United States, English departments taught and studied writing—thus, they became the homes of writing programs—yet both pedagogy and application were largely focused on analysis of great literary texts. As they turned to the challenge of understanding people's documentation needs, this background turned out not to be sufficient. The works of Chaucer, Melville, and Hemingway turned out not to be strong models for writing instructional documentation. So, as professional communication programs attracted people with no background or interest in literary analysis—like me, a computer science major—they also had to seek different theoretical and empirical frameworks for understanding the activities in which users were involved. They essentially had to start with a blank page, seeking a framework that could guide *qualitative research* to *describe work activities* mediated by *multiple texts*. In workplace studies, these often resulted in *new texts and textual systems* for *situated audiences*.

To better understand how CHAT was taken up in writing studies, we'll again examine domain, project, translations, and trials.

- **Domain**: The domain was writing studies. Scholars in this field translated Engeström's CHAT by bringing it into contact with other discourses addressing how people use texts in practical activity.
- **Project**: Writing studies' project was effectively describing and diagnosing types of *mediatory texts*, often working in *systems*, that could be mapped so that the dynamics of specific workplaces could be better understood

and addressed. In the majority of cases, this project involved acting with technology: understanding how people communicate, coordinate, and collaborate at work; mapping out the types of texts (especially digital technologies) that mediate work; and identifying ways to improve how those text types were constructed, used, and coordinated.

- **Translations:** To produce appropriate descriptive research, writing studies scholars translated Engeström's CHAT in several ways. We'll specifically examine translations in empirical focus, mastery, and mediators. Most significantly, Engeström's CHAT was synthesized with genre theory, and CHAT studies were implemented as small-scale case studies. The sum of these translations was a lightly modified version of Engeström's CHAT, focused on conducting and describing small-scale case studies with multiple mediators.
- **Trials:** Finally, this settlement also had to weather trials. These trials included both theoretical and methodological challenges.

In this chapter, I first provide a historical account of how North American writing studies developed in the twentieth century and how Engeström's CHAT offered this field a sociocultural base. Next, I examine this iteration in terms of the three translations mentioned—empirical focus, mastery, and mediation—and how these translations led writing studies to address their project of mapping and diagnosing mediatory texts. I conclude by identifying trials that tested CHAT as a sociocultural base—trials that are arguably ongoing in writing studies but that have had limited impact on broader CHAT discourse.

AN EXTRAORDINARY CASE: A STICKY NOTE

In 1998, just over a decade after Engeström and Bødker published the pivotal works described in chapter 6, I visited a police station in Des Moines, Iowa. I was there to watch Barbara, a police officer, determine where to set *speed traps*: stretches of road that patrol cars would monitor to identify speeding and other moving violations. To determine these locations, Barbara used a database called PC-ALAS to identify areas with high numbers of traffic accidents related to moving violations. This database dated back to the 1960s (Iowa was the first state in the country to computerize its traffic accident data), and it was the focus of my dissertation research (published later in Spinuzzi, 2003b).

PC-ALAS had a text-only interface, and its underlying database was exceedingly simple. Various road features (such as intersections and highway on-ramps) were identified by a six-digit code, not by an absolute location. When a user wanted to examine a specific intersection, they would have to look up the intersection's six-digit number on a specialized map: about a meter square, rolled up and stored in a cardboard tube. To examine the map, they would clear a desk, roll out the map, and place objects (coffee cups, paperweights, books) on the corners so that it would stay flat. Then they would find the intersection that they wanted to investigate, and they would read the six-digit number. Then they would turn to a different desk, where the work computer sat, and type in the six-digit number. I had visited several offices across Iowa that used PC-ALAS—offices of city and county engineers, police officers, and grantwriters writing grants for Federal Highway Administration funds—and this moment frequently introduced transposition errors: in turning from one desk to another, the user often transposed two digits in the six-digit number they kept in their memory. Sometimes they would catch the error when they saw the database output. Often they wouldn't.

But Barbara had no such transposition errors. At some point in the past, she had simply written down the numbers of the intersections on a sticky note, which she kept in a folder. When it came time to query the database again, she simply took out this sticky note and put it on the frame of her monitor, then typed the number into her computer.

Barbara's solution was mundane. Certainly, she thought it was unremarkable, and she reacted with polite disinterest when I asked her about this solution, which she thought was obvious. Yet the effect of this minor addition on her work was extraordinary: it meant that she did not have to deal with the frustrations and errors that other PC-ALAS users had. She spent less time on the task, and her results were more reliable, not because she was more skilled, but because she had introduced a new type of text (or *genre*) that offloaded a difficult task onto a simple, easy-to-use type of text. And she used this text type in coordination with a relatively stable set (or *genre ecology*) of other text types, such as interface features and the map; these texts *comediated* the activity through specific connections. I could see Barbara making these connections, and I could ask her to reflect on these connections in my interview with her

In fact, as I observed her and other users, I realized that *all* of them were comediating their work activities with genre ecologies. Some of these genres were "official," supplied by the Iowa Department of Transportation, but many others were "unofficial" innovations implemented by individual users, such as the sticky note. Since these users had no mechanism to interact with each other, such unofficial innovations did not spread.

My observations and interviews with PC-ALAS users were not unique. At about the same time, others in writing studies were conducting qualitative case studies of professionals using sets of text types to perform work, and many were beginning to use Vygotskian theory and Engeström's CHAT to guide these studies (e.g., Berkenkotter & Huckin, 1995; Haas, 1996, 1999; Winsor, 1999; Dias et al., 1999; Artemeva & Freedman, 2001). These investigations used the concept of genre to conceptualize how these professionals read, understood, used, and composed these text types, and many also used CHAT to conceptualize the social settings within which this work happened. Through the 1990s and 2000s, genre theory and CHAT were synthesized to provide an investigatory framework that was well suited for conducting descriptive qualitative case studies of workplace writing.

In the PC-ALAS case and similar ones, extraordinary feats were understood as being made possible not through single mediators but through interactions among types of textual mediators: both mediators passed down to participants and mediators consciously developed or adapted by those participants. Through genre theory, each mediator was seen as emerging from recurrent responses to rhetorical situations, and the interactions among these mediators were seen as meetings of different cultural heritages, domains, logics, and responses. Everyone knew how to read a map, but not everyone was prepared to understand the six-digit note numbers superimposed on a node map, nor were they always able to interpret the interface's dialog boxes—which themselves were representations of older printed forms coming from different domains. And concrete observations made it obvious that users supplemented the given genres with their own genres, imported from elsewhere, creating further clashes among mediators. Whereas Bødker and Engeström told stories of interventionist research, in which participants focused on *codesigning interfaces and work practices*, case studies of workplace writing described stories of participants who had themselves *supplemented interfaces and rewired work practices* by enlisting and modifying additional mediators.

How did we get here? How did we see ourselves as faithfully enacting Engeström's CHAT when, in retrospect, we were using a different research approach and emphasizing interactions among mediators?

HISTORY

Writing is one of our most protean tools, used across curricula and professions, but also in many facets of our lives. In the United States, as in most of the developed world, the population is over 99% literate and writing is everywhere.[1] Our basic infrastructure, our locations, even our food and clothing are labeled with texts. Yet college-level *writing instruction* has been traditionally treated as the domain of English departments, typically under the heading of "composition."

A BRIEF HISTORY OF WRITING STUDIES IN ANGLOPHONE NORTH AMERICA

To understand how CHAT was taken up in North American writing studies, we must understand the idiosyncratic history of writing as a focus of academic specialization in anglophone North America. As David R. Russell (1997b) explains, North American writing instruction involves "three activities that are virtually unique to North America: general composition courses, writing-across-the-curriculum (WAC) programs, and business and technical writing courses" (p. 225). All three are associated with English departments, and all were to some extent marginalized through most of their development.

COMPOSITION Russell summarizes:

> General composition courses, usually housed in English departments, have been the single curricular common denominator in U.S. higher education since the rise of the modern university. . . . The motive of these courses is to improve students' writing and prepare them for academic and workplace writing, which was assumed to be a single, generalizable skill learned once and for all, usually at an early age. Throughout their history, 'Freshman Composition' courses have been considered remedial, have performed a primary gatekeeping function in many universities, and have been considered marginal to the central object of English departments—literary study—which was generally conceived in formalist terms. Students have been given a set of precepts and some models, then told to write—well. (Russell, 1997b, p. 225)

WRITING ACROSS THE CURRICULUM (WAC) He continues:

> In the late 1970s, however, composition teachers professionalized, developing several strands of research drawing from both the humanities (i.e., classical rhetoric) and the social sciences. The social scientific strands were initially based on behaviorism, information-processing cognitive psychology, and Piaget's structuralism. Researchers examined the psychology of individual writers, attempting to find a

> generalized "writing process" individuals use. However . . . cross-disciplinary rhetorical comparisons made the limitations of these generalized approaches apparent with the rise of the WAC movement in the early 1980s, which grew out of the newly professionalized field of composition (Russell, 1991). In response to the influx of previously excluded groups with open admissions policies in higher education, many colleges and universities began programs to enlist the aid of faculty in all disciplines to improve students' writing. As teachers and researchers of composition interacted with faculty in other disciplines, through workshops and consulting, they glimpsed the great variety and richness of the uses of writing and began to rethink fundamental assumptions that have undergirded general composition courses for a century. (Russell, 1997b, p. 225)

BUSINESS AND TECHNICAL WRITING Russell summarizes the state of business and technical writing research:

> University-level business and technical communications courses have been taught in North America since the 1st decade of the 20th century. Generally housed in English departments and, like composition, generally marginalized, these courses have also traditionally taken a formalist approach, with students learning proper formats for letters, memos, reports, and so on. With the publication of Odell and Goswami's (1985) *Writing in Non-Academic Settings*, ethnographic methods were applied to questions of writing in the activities of professionals. Here again the goal was to improve the teaching of writing by learning about how writing mediates the actions of workplace professionals. Writing research moved beyond academia and into other activity systems, where writing researchers and consultants brought their expertise to bear on a wide range of problems. (Russell, 1997b, pp. 225–226)

Although these activities were separate, they typically drew from the same general pool of people, which we'll call "compositionists." Russell himself is one example: after being trained in literary study, he began teaching composition, then studied and wrote about WAC, and then edited the flagship *Journal of Business and Technical Communication*. So it is with hard-won personal experience that Russell relates the history of all three.

COMPOSITION BECOMES A RESEARCH FIELD

As Russell says, through most of the twentieth century, composition in the United States was generally not considered to be a field of study, but rather an uninteresting pedagogical specialization. In 1971, James Kinneavy complained, "Composition is so clearly the stepchild of the English department that it is not a legitimate area of concern in graduate studies, is not even recognized as a subdivision of English in a recent manifesto put out by the major professional association (MLA) of college English teachers . . . in some universities is not a valid area of scholarship for advancement in rank, and is generally the teaching province of graduate assistants or fringe members of the department" (Kinneavy, 1971, p. 1). He added, "There is no definite concept of what the basic foundations of composition are" (p. 2).

By 1985, composition had gained ground, but Maxine Hairston argued that compositionists remained limited by their colleagues in literature, their "intimate enemy." Hairston exhorted us to develop our own theories and conduct our own research. We thus needed "guidelines to define and govern good research," she argued, and should "extend our connections to disciplines outside our field . . . by establishing connections and credibility outside of English departments, we stand to improve our standing with the whole university community" (Hairston, 1985, pp. 279–280). To emerge as a field, composition needed a research orientation. Composition had a broad research object (writing), but it also needed a *paradigm* within which to understand its research object and the boundaries of its investigation; *methodologies* to underlie, motivate, and guide research; *methods* through which researchers could investigate phenomena consistently; and *techniques* for implementing those methods (Spinuzzi, 2003b, 2005b).

Yet composition scholars did not agree on what this research field should look like. Several research agendas sprang up, drawing from different bodies of knowledge but without an agreed-upon paradigm for validating and connecting research (North, 1987). Faigley (1986), in one of many framework essays appearing in the late 1980s to early 1990s (see also Berlin, 1988; Blyler & Thralls, 1993; Nystrand et al., 1993), summed up three competing camps:

- **Expressivists,** who focused on individuals' authentic expression, had a comparatively thin, anecdotal research tradition.
- **Cognitivists** systematically investigated and compared individuals' writing processes, drawing on the paradigm of information-processing cognitive psychology; they used experiments and think-aloud protocols to map out common processes using both quantitative and qualitative measures.
- Finally, **social constructionists** applied the proposition that "human language (including writing) can be understood only from the perspective of a society rather than a single individual" (Faigley, 1986, p. 535); they investigated social context by drawing from "poststructuralist theories of language, the sociology of science, ethnography, and Marxism" (p. 535). Empirically, they drew from sociology and anthropology, using primarily

qualitative measures. Exchanges among these camps grew more heated (Berkenkotter, 1991) leading to "sweeping generalizations about the character or ethos of researchers in general" (Charney, 1998, p. 28).

Meanwhile, research method textbooks for composition studies tended to be paradigm-agnostic, presenting different methodologies as legitimate but paradigmatically separate ways to investigate research questions (Kirsch & Sullivan, 1992; Lauer & Asher, 1988). The methodologies sat uneasily next to each other, barely interacting, like shy teens at a dance. Some studies published at the time were mixed-method studies, but the research textbooks generally did not cover mixed-method research in enough detail to execute it well.

This state of affairs was problematic, especially for those like me, who came into writing studies in the early 1990s with an interest in professional communication (an umbrella term for what Russell (1997b) calls "business and technical writing"). Personal computers had quickly become a fixture at offices, with far-reaching implications not just for how we worked but also for how we organized work. Like the Danish typesetters that Bodker studied in the 1980s, US professionals in the 1990s found that their work was radically changed by new hardware and software. Unfortunately, the documentation and interfaces that shipped with these new technologies were often not up to the task, focusing on describing features rather than helping users to relate these features to their own tasks and lives (cf. Carroll, 1990; Haas, 1996; Coney & Chatfield, 1996). As an undergraduate computer science major and technical writing minor in the late 1980s and very early 1990s, I had helped many people as they struggled to understand their hardware and software, and I had seen the negative impact of their struggles—not just on their quality of work output, but more importantly, on their quality of working life.

How were we to study such work, and how could we guide positive changes based on such studies? As a new Ph.D. student in 1994, I read the framework essays and the exchanges in our field's journals, and I felt torn. On the one hand, cognitivists offered a focus on individual development, empirical meaning-making, and comparisons across populations. They offered an avenue for rigorously examining how people work with texts, conceptualizing their cognitive processes, and designing information resources that could work in concert with those cognitive processes. And they offered methods such as quasi-experiments, field experiments, and think-aloud protocols. On the other hand, social constructionists accounted for the indeterminacy of interpretation and the contingent, cultural context of meaning-making. They offered ecologically valid ways of investigating how people work in groups, and they offered methods such as qualitative case studies and ethnography.

Both social and cognitive approaches had something to offer. And at about the time I entered my Ph.D. studies, compositionists had begun exploring *sociocognitive* frameworks. For instance, Linda Flower (1994) advocated for a "social cognitive theory of writing" based on Vygotsky (cf. Flower, 1979). Geisler (1991) proposed exploring mental models to develop a "sociocognitive model of literacy," drawing on Scribner & Cole (1981) among others. Similarly, Charles Bazerman (1988) drew on sources such as Merton's work in the sociology of science and Carolyn Miller's (1984) work in genre theory to theorize genre as a "sociopsychological category" (p. 319). Dorothy Winsor (1994, 1996, 1998) drew on a variety of approaches such as situated and distributed cognition (both of which offered what could be called sociocognitive understandings of human activity), actor-network theory, and institutional ethnography; the latter was also used by Catherine Schryer (1993) in her studies of the genre of medical records.

CHAT ENTERS THE FIELD AS A SOCIOCOGNITIVE APPROACH

I can't confidently map out all vectors by which Engeström's CHAT entered writing studies. But I can confidently map one specific vector. Bazerman reports first hearing about Engeström's work at a conference in 1992 (2023, p. 144, footnote 7). He subsequently obtained a copy of Engeström's *Learning, Working, and Imagining* (1990), which he heavily marked up. Bazerman then copied the book for David R. Russell, who became my dissertation supervisor, and David allowed me to make my own copy (which I still have, with Bazerman's name written across the top of the title page). David had already cited Vygotsky in earlier work (1993), and his first discussion of Engeström's CHAT soon followed (1995). By 1997, several studies in our field had used CHAT in some form (Russell, 1997b). Many of these early works characterized CHAT mainly in terms of other sociocognitive approaches: as a synonym for situated learning (Blakeslee, 1997) or practice theory (Haas, 1996) or as important for its influence on situated cognition (Berkenkotter & Huckin, 1995). It was often used in coordination with communities of practice theory, distributed cognition, and semiotic theory (Dias et al., 1999).

In theorizing and investigating cognition as thoroughly social, CHAT offered a solution to our tangled methodological problem. First, it addressed the social-cognitive divide via a dialectical synthesis. Second, since it accounted for individual and social development, it could underpin writing research across various contexts. Third, it allowed us to integrate methodologies used by both cognitivist and social constructionist researchers: a CHAT research design could investigate meaning-making across mixed methods. That meant that, fourth, it could allow us to reinterpret and integrate previous work across paradigms.

More granularly, Engeström's CHAT offered a few other things that made it attractive for composition scholars, who largely came to the field via English departments and thus did not have deep grounding in empirical approaches:

- **A handy pocket manual**: Engeström's triangle (and the categories named in it) functioned as a "handy pocket manual of social analysis" (Blunden, 2010, p. 232). Although Blunden does not mean this as a compliment, the heuristic was a huge boon for those of us with little social science training. Whether we drew literal triangles or not—some of us drew circles or hexagons, or just used textual description—we had a conceptual language for describing the different cases we investigated. Some of us even used the triangle as a pedagogical tool (e.g., Kain & Wardle, 2005).
- **Models for descriptivist research**: The small-scale case studies in *Learning, Working, and Imagining* (Engeström, 1990) and *Interactive Expertise* (Engeström, 1992), although mainly descriptive rather than interventionist (atypical in Engeström's work), provided workable models for those who were interested in case studies of workplace writing. Especially interesting were discussions of multiple texts being used in these cases, "alternative historically grounded representations which can be understood and appreciated against the background of their cultural evolution. . . . [manifested] on the actual choice and use of tools," such as a computer terminal on which a doctor has pasted handwritten notes (Engeström, 1990, p. 177). We recognized the potential for examining writing as mediating concrete activities.
- **A link to genre**: Engeström's emphasis on dialogism in *Learning by Expanding*, and on mediating texts in *Learning, Working, and Imagining*,

was followed by Ritva Engeström's (1995) discussion of voice as communicative action, which interpreted Bakhtin's dialogism via Leontiev's levels of activity (applied to writing studies in Russell, 1997a, 1997b and Spinuzzi, 2002b). We saw the potential in connecting CHAT with Bakhtinian genre theory, which we had already embraced.

ENGESTRÖMIAN CHAT IS SYNTHESIZED WITH GENRE THEORY

Whereas Engeström was interested in expertise and Bødker was interested in designing computer interfaces, those of us in writing studies were more interested in how people produce and use texts. Leontiev had characterized a tool as "a material object in which are crystallized methods and operations, and not actions or goals" (Leontyev, 1977/2009a, p. 102). But, drawing on theorists such as Bitzer and Bakhtin, we understood texts as more than bundles of operations. Rather, we understood text types as functioning as *genres*, specifically, as repeated responses to repeated rhetorical situations (Miller, 1984). That is, texts had to be understood as related to some repeated activity, iterated and developing over time, sedimenting into specific forms that amount to forms of life (Bazerman, 1988). Bazerman provides an eloquent illustration in his examination of how the genre of the experimental article developed over time, starting as "simply cookbook recipes for creating marvellous effects or effects of practical use, such as the directions for coloring marble internally" (p. 66), and progressing to the comparatively rigid set of categories we associate with the experimental article today. In this example, the genre of the experimental article develops significantly, but this development does not simply incorporate operations; it provides a conceptual structure that new practitioners learn as they approach their own experiments. Furthermore, Bazerman argues, the genre continues to change.

As we can see here, a single text is not the same as a genre: as a repeated response to repeated rhetorical situations, a genre is a *type*, a pattern. Studying a pattern requires historical comparisons (how has the pattern *developed*?) as well as a relationship to some sort of framing activity (how does the patterned text *respond* to a repeated rhetorical situation?). Without those two aspects, it's very difficult to interpret a text. For example, if we examine the text of a Bible, we might draw many conclusions about how it's used. But if we walk into someone's office, we might find that they put a thick Bible under their computer monitor to raise it to eye level. In practice, texts are sometimes abducted—and modified and juxtaposed with other texts.

This brings us to a second insight. Although one genre could be isolated for study in the way Bazerman studies the experimental article, genres typically relate to other genres as well as the framing activity. Bazerman (1994) calls these interrelationships *genre systems* and gives the example of interrelationships between patent applications and patent grants. As he argues later, "Texts do not appear in isolation, but in relation to other texts" (Bazerman, 2004a, p. 53). This fact is clear enough even in historical studies but becomes even clearer in qualitative studies, in which an observer can readily see how professionals gather, assemble, and interrelate texts—texts such as the PC-ALAS interface, node map, and sticky note mentioned at the beginning of this chapter. Like Bazerman, others began developing accounts of how genres interrelated in sets, systems, repertoires, ecologies (cf. Spinuzzi, 2004), chains, colonies, and suites (cf. Andersen et al., 2014). These accounts sometimes drew on Hutchins' (1995) description of tool ecologies (especially Spinuzzi, 2003b).

Hence, each genre has a *developmental history* (such as PC-ALAS's dialog boxes, which developed from older paper forms), an *orientation to the current activity* (the dialog boxes generate specialized reports for specific types of users), and a *relationship with other genres* being used within that activity (the dialog boxes relate to both the reports and the node maps). These genre relationships meant that genres comediated the activity (cf. Bødker & Andersen, 2005, who drew similar conclusions). Since individual users could and did enlist additional genres in their work, that comediation could look different for each individual. And since genres have their own developmental histories and preexisting orientations, enlisting new genres could involve dragging in those histories and orientations, potentially destabilizing the activity under consideration.

To better theorize the developmental angle and the orientation angle, genre scholars took up CHAT. Russell (1997b) overviewed early research along these lines, while another article he published the same year (1997a) provided a detailed theoretical discussion of Engeströmian CHAT and its relationship to genre and genre systems. This genre-CHAT synthesis soon underpinned observational case studies of workplaces (e.g., Winsor, 1999; Haas, 1999; Dias et al., 1999; Artemeva & Freedman, 2001; Bazerman & Russell, 2003; Spinuzzi, 2003b; Maistre & Pare, 2004), and CHAT was named as one of the "foundations for research" on "texts mediated by information technologies" (Geisler et al., 2001). The latter half of the 2000s saw a glut of such research (e.g., Schryer & Spoel, 2005; Smart, 2006; Spafford et al., 2006; Schryer et al., 2007; Spinuzzi, 2007; Winsor, 2007; Artemeva & Freedman, 2007; Little, 2007; Spinuzzi, 2008; Sherlock, 2009), some of which began to incorporate networks of activity as a way to describe organizations. Meanwhile, the genre-CHAT synthesis continued to be theorized (Bazerman et al., 2003; Bazerman, 2004b; Artemeva, 2008).

Critically, although writing studies researchers were aware of Engeström's later work (discussed in subsequent chapters), and to some extent acknowledged some of the concepts of this later work, such as knotworking, CHAT work in writing studies remained largely rooted in Engeström's CHAT of the 1980s and early 1990s, both theoretically and methodologically. We'll see why below, as we examine how writing studies treated empirical focus, mastery, and mediation.

EMPIRICAL FOCUS

Given the focus on genres as "stabilized-for-now" (Schryer, 1993) patterns of texts in use, and on their intertextual relationships with each other, writing studies researchers mainly conducted workplace studies as descriptive qualitative case studies in small scenes. These small scenes allowed us to focus on complex interactions among genres in those small scenes.

These scenes included individuals working with text sets (e.g., Spinuzzi, 2003b; Sun, 2012; McNely, 2017), small collaborative groups and teams (e.g., McCarthy et al., 2011; Opel & Hart-Davidson, 2019), and departments, offices, units, organizations, and cross-organizational collaborations (e.g., Artemeva & Freedman, 2001; Spinuzzi, 2008, 2015; Maistre & Paré, 2004; Gygi & Zachry, 2010; Fraiberg, 2017). These were typically structured using case study methodology, in which the activity system was interpreted as providing a case boundary (see Spinuzzi, 2023b). In being bounded, these case studies allowed writing studies researchers to focus on complex mediatory interactions among genres as well as transformations within these genres.

To describe relationships across cases, such as interacting departments and organizations, some researchers turned to networks of activity (esp. Spinuzzi, 2008). As mentioned in the previous chapter, networks of activity provided a way to map organizations. For instance, Winsor characterized complex organizations (such as corporations) as activity networks composed of related activity systems (such as work teams), operationalized as linked cases (1999,

p. 201), and she characterized a corporation and a labor union as "competing activity networks" (p. 217).

This pragmatic measure, however, creates an issue in the CHAT understanding of division of labor. Leontyev (1959/2009b, p. 186) describes fire-tending as an *action* that makes sense only in the context of the larger activity of the hunt. But in describing networks of activity, Engeström made it possible to describe specialized pursuits (such as fire-tending) as *activities in themselves*, activities that are networked to other activities. Carrying that innovation forward to our case studies of writing arguably meant that we fractured the concept of activity, delineating separate contexts or cases that meet only through the components of their activity systems (but compare Russell, 1997a). At the same time, applying activity systems in this way allowed us to bound and scale our empirical cases in such a way as to support detailed study of texts (for examples of detailed lists of text mediators, see inter alia, Haas, 1999; Spinuzzi, 2003; Winsor, 1999). That is, it gave us a bounding principle for case studies, one that allowed us to draw on case study methodology (see Spinuzzi, 2023b) so that we could manage data collection and analysis in small scenes of activity (see chapter 10 for more discussion on this point).

These small scenes or cases contrast with the broader-scale interventionist research that Engeström was beginning to conduct during the same period. As we'll see in the next chapter, as Engeström moved to broader scales, it made less sense to practically attend to the details of the many mediators being used at larger scale. While we in writing studies focused on the complex assemblages of genres at play in bounded cases, Engeström and his collaborators looked past those detailed assemblages, focusing instead on codesigning coordinative genres that could reorganize the other texts: *agreements* such as care agreements, mobility agreements, and rental agreements. (We'll discuss these agreements in chapters 7 and 8.)

MASTERY

We've seen that in previous implementations of CHAT, foci included self-mastery, state mastery, and the collective mastery of participants deliberating over instruments and the division of labor. In writing studies, however, the focus tended to be on describing how participants mastered specific genres and the interconnections among them.

In terms of *mastering specific genres*, case studies included students who learned to take on professional identities by mastering a pivotal genre (e.g., Doody & Artemeva, 2021; McNair & Paretti, 2010; Schryer et al., 2007; Schryer & Spoel, 2005; Spafford et al., 2006), as well as how professionals enacted and changed a genre at work (e.g., Artemeva, 2005; Haas, 1999; Propen & Schuster, 2010; Schryer et al., 2009; Schuster & Propen, 2011; Spafford et al., 2010; Varpio et al., 2007, 2008). In many cases, mastering a genre was associated with taking on a professional identity and asserting professional agency (e.g., Artemeva, 2005, 2009, 2019; Artemeva & Freedman, 2001; Winsor, 1999, 2000, 2001). Some of these studies foregrounded a specific genre while acknowledging its interconnections with others.

In terms of *mastering interconnections among genres*, case studies mainly focused on professional settings in which participants had to conceptualize genre assemblages and their relationships (e.g., Ding, 2008; Gygi & Zachry, 2010; Smart, 2006; Wall & Spinuzzi, 2018). These often included cases in which participants added or modified critical genres to address problems in the ongoing work (e.g., Bazerman et al., 2003; Hashimov & McNely, 2012; Spinuzzi, 2003b, 2008, 2015a). And they sometimes involved cases in which introducing a novel genre involved reorganizing the relationships among other genres (e.g., McCarthy et al., 2011; Sherlock, 2009; Spinuzzi, 2003b).

Since genres were conceived as rhetorical responses to recurrent rhetorical situations, learning a genre was understood as tapping into the cultural heritage of those who had come before, and importing a genre from elsewhere was interpreted as an agentive act that tied one cultural heritage to another—especially in the face of rapid changes wrought by the ongoing digitalization of workplaces and the resulting changes to work organization (e.g., Artemeva & Freedman, 2001). Broadening the repertoire of genres meant broadening the range of cultural heritages one could draw upon when solving complex problems. And since the understanding of genre in writing studies was heavily influenced by Bakhtinian perspectives, these cultural heritages were largely understood as dialogical, with meaning emerging from their differences, differences that do not need to be dialectically reconciled. As we'll discuss in chapter 11, this dialogical understanding of mastery is distinct from the ultimately dialectical understanding of mastery in Engeström's CHAT.

MEDIATION

This brings us to mediation. In chapter 6, we saw that Engeström and Bødker focused on cooperatively designing mediators to be used in a target activity: a computer interface or a division of labor. But in writing studies, the focus was on how people produced and used *types* of texts (i.e., genres). The distinction is important because in writing studies, the participants routinely produce unique texts in recognizable patterns, recognize those patterns in others' unique texts, and relate assemblages of texts based on those recognizable patterns. This is true even when participants do not consider themselves to be writers (e.g., Spinuzzi, 2015a). Whereas Engeström and Bødker staged intervention events that produced relatively static mediators that could move work from one relatively stable state to another, in writing studies, we mainly described scenes of work in which written mediators were produced and interrelated on an ongoing basis. Some of these genres of written mediators were common and collective, understood and regulated by the organization. But many were idiosyncratic, being improvised or imported by an individual from another activity to solve their distinct problem and then sometimes spreading to others.

It's worth emphasizing that in being oriented to both consumption and production, genre represents a break from the tool perspective that was bequeathed to CHAT by Leontiev and that anchored Bødker's human-computer interaction (HCI) work.

According to Leontiev, "a tool is a material object in which are crystallized methods and operations, and not actions or goals" (2009a, p. 102). We can, for instance, consider how a line of craftspeople progressively craft axes that incorporate more and more work operations, resulting in axes that are more and more adapted for a specific sort of action. When an axe is taken up, the user uses it as intended: it is "a socially developed means of action, namely the labour operations that have been given material shape, are crystallised, as it were, in it. That is why to handle a tool means not simply to hold it but also to know how to use it" (2009b, p. 192). In applying Leontiev, Bødker understands design as a process of learning (Bødker, 1987, p. 45), a process in which new operations are built into the tool to set the new conditions for collective action (p. 46). That relatively static tool is then repeatedly used by different individuals, just as an axe is used by individuals who learn how to handle it properly.

In contrast, genre is understood as a *pattern* that is produced and interpreted by people in the course of the activity. Genres such as monthly reports (Spinuzzi, 2015a), victim impact statements (Propen & Schuster, 2010), and electronic medical records (Opel & Hart-Davidson, 2019) thus develop rapidly and sometimes idiosyncratically, can be abducted into adjacent activities and repurposed, and often serve as boundary objects that are used and interpreted differently by intersecting activities (Popham, 2005). Whereas Leontiev's tool is produced by a designer who crystallizes operations that set the conditions for users' proper tool handling, genres are collectively produced through their repeated use by users. Leontiev's tools are produced *for* mediation, while genres are often produced *during* mediation.

Finally, as we've seen, genres are often understood as mediating activity in concert with many other genres. To take the example at the beginning of this chapter, the sticky note's importance is not as a supermediator, a consciously redesigned artifact that was designed to revolutionize the activity. Rather, the sticky note is an addition to the other genres being used to collectively mediate the activity, bridging their many cultural heritages. This compound mediation has been explored in many genre studies, not just in those using CHAT (see Andersen et al., 2014 for a review), and it implies a constant (and often constantly changing) dialogic tension among different genres with their different cultural heritages.

SUMMARY: TRANSLATING CHAT TO ACHIEVE A SOCIOCULTURAL ACCOUNT OF WRITING

In the early 1990s, writing studies scholars saw CHAT as offering one possible base for a sociocognitive approach to writing studies, one that could offer a theoretical and methodological base for understanding writing in context, especially in workplaces. CHAT was attractive in part because it was compatible—or at least compatible enough—with genre theory, which had already begun to be used in studies of writing. Although CHAT was not universally taken up, it became an important and accepted framework and underpinned much work, especially in workplace writing studies. One measure of this success is that CHAT-specific terms such as "activity," "activity system," and "activity network" are now often used without citations in writing studies; the terms are simply regarded as part of the vocabulary of the field.

In taking up CHAT, writing studies translated it for the field's localized concerns. These translations made CHAT more useful for writing studies:

- It became oriented to genre, focusing on the production and interpretation of text types rather than designed mediators in organizations.
- It became descriptive, underpinning case studies rather than interventionist studies, with participants portrayed as gaining and building their own agency by mastering and improvising genres.
- It examined mediation in terms of genres and interrelations among these genres, necessarily keeping case studies on a small scale in order to examine these interrelations.
- In focusing on genre, it balanced dialectical understandings of activity with dialogical understandings of genre: interrelated genres were understood as maintaining tensions among different cultural heritages and different legitimate discourses, and these tensions did not necessarily need to be reconciled.

These changes can be summed up in three different translations:

- **Empirical focus**: In this localized translation, CHAT's empirical focus shifted from deliberations about labor activity in larger groups to closely bounded descriptive case studies of work.
- **Mastery**: Its understanding of mastery shifted from participating in the design of mediators to producing and interpreting genres and genre assemblages.
- **Mediators**: Its mediators themselves were no longer designed in an arranged design activity but, instead, were genre instances that often bridged activities and interacted with other genres.

Through these translations, writing studies scholars

- **problematized** existing cognitive and social accounts in writing studies;
- **interested** writing studies researchers as well as those who produced and used genres in professional education and workplaces;
- **enrolled** CHAT figures from other fields and disciplines (e.g., Vygotsky, Leontiev, Engeström, Bødker, Nardi, and many others), genre theorists (e.g., Bakhtin, Miller), and qualitative research methodologists;
- **mobilized** these enrolled actors to carry out descriptive case studies research in classrooms and workplaces, focusing on the details of the interrelated text types involved in these cases.

Through their translation of empirical focus, mastery, and mediators, writing studies scholars produced a theoretical and empirical basis

Table 7.1
Domain, project, and translations in writing studies' uptake of CHAT

Domain	Writing studies
Project	A theoretical and methodological base for understanding professional writing, especially in workplaces
Translations	
• ***Empirical focus***	Closely bounded descriptive case studies of work
• ***Mastery***	Producing and interpreting genres and genre assemblages
• ***Mediators***	Genre instances that often bridged activities and interacted with other genres

for writing-oriented case studies. This settlement was very different from Engeström's 1987 settlement. It was also largely confined to writing studies, with very limited uptake elsewhere. For that reason, we'll omit the discussion of trials in table 7.1.

In this interlude chapter, we've taken a side trip to closely examine one localized uptake of CHAT for specific purposes. Yet this chapter is an interlude precisely because this localized uptake has not significantly impacted Engeströmian CHAT beyond this field. So, in the next chapter, we rejoin that main strand of CHAT development, examining how Engeström and collaborators transformed CHAT into a de facto organizational sociology.

8 ENGESTRÖM'S "THIRD-GENERATION" CHAT: AN ORGANIZATIONAL SOCIOLOGY

As we rejoin the mainstream of what we can now call "Engeström's CHAT," let's take a moment to consider how he used his triangles and other visuals in his early work. As we saw in chapter 6, these visuals played a double role.

On one hand, Engeström used these triangles as a practical shared representation that he could use with participants in interventionist projects: people who were untutored in CHAT or social theory could examine a triangle and envision the social system in which they worked, and with some prompting, they could also envision changes to that social system. In this sense, they were like Ehn's language games or other practical visuals used in design activities (e.g., Beyer & Holtzblatt, 1998; Di Russo, 2016; Griffiths et al., 1998).

On the other hand, Engeström also used these triangles when explaining activity theory to researchers and theorists. In these publications, he positions his interventionist studies as a way to "*illuminate* the theoretical arguments" (2000, p. 960, my emphasis) and to "*concretize* the theoretical ideas" (2001, p. 133, my emphasis). These publications describe and promote the theoretical project, illustrating it with his interventionist studies. As he tells one interviewer, "Take no shortcuts, this is not a superficial fad you can pick and use as theoretical decoration in your studies. This approach is demanding both theoretically and practically" (Engeström & Glăveanu, 2012, p. 518).

This interplay between practical and theoretical agendas is a key dynamic in Engeström's CHAT, particularly in the mid-1990s as CHAT was applied to a broader scope.

In developing CHAT into an approach for collectively describing, modeling, and intervening in organizations, Engeström realized early success but also found limitations. Some of these cases were difficult to fit into activity systems and networks, specifically cases with (1) larger spatial and temporal scope and (2) different types of work. Thus, the mid-1990s began a sustained period of theoretical development for Engeström and collaborators, centered around the spatial and temporal expansion of the object. Closely bounded scenes gave way to organizational and cross-organizational activity over longer periods of time and across different settings, and simultaneously, Engeström's CHAT drew from and was positioned alongside "anthropology, psychology, communications, sociology, sociology of science, and cognitive science" (Engeström & Middleton 1996, p. 1). The result was a de facto *organizational sociology*, one that specifically addressed the development, structure, and functioning of organizations and interorganizational activity. Engeström's CHAT in this phase still focused on acting with technology but emphasized *acting* over *technology*—a necessary shift in emphasis, since the expanded scope of investigation meant that the investigated activity was mediated by a whole set of interacting mediational resources (cf. Bødker & Andersen, 2005).

This organizational sociology mainly emerged from the work of Engeström and collaborators under his direction at the Center for Research on Activity, Development, and Learning (CRADLE, established in 1994) at the University of Helsinki. During this period, much of the work was funded by the Academy of Finland, the Finnish Work Environment Fund, and the Finnish Funding Agency for Technology and Innovation (TEKES), and research was conducted with partners such as the City of Helsinki Health Centre. In concert with this institutional funding, the research expanded from dyads and small groups to institutional scope.[1]

To better understand Engeström's CHAT in this period, let's again examine domain, project, translations, and trials.

- **Domain**: The domain that Engeström addressed in this period was organizational sociology. Engeström and colleagues translated insights from Westernized CHAT to address how organizations worked and how they could be made to work better.
- **Project**: Engeström's project—what he was trying to achieve—was the reorganization of interorganizational work. This overarching project, which we'll discuss below, drove Engeström's attempt to develop his organizational sociology.

- **Translations:** To produce an organizational sociology for reorganizing interorganizational work, Engeström and colleagues translated CHAT in several ways, largely by putting it in conversation with existing concepts in managerial and social theory, including concepts from Barley, Deleuze and Guattari, Giddens, and Victor and Boynton. These insights were taken up within the CHAT framework to develop a de facto organizational sociology that could be deployed to generate managerial insights and support collective change. We'll specifically examine translations in empirical focus, mastery, and mediators. The sum of these translations was his new settlement, a cultural-historical organizational sociology.
- **Trials:** Finally, this settlement had to weather trials, including skepticism from managerial and social theorists as well as internal skepticism from others in CHAT.

Let's look at one case that exemplifies this sociological approach, and then we'll analyze this approach's history and its treatment of empirical focus, mastery, and mediators.

AN EXTRAORDINARY CASE: APPREHENDING A MENTAL PATIENT

In the mid-1990s, Engeström and CRADLE collaborators began to question the limits of the team-based models they had helped to develop in Finnish medical clinics (Engeström et al., 1999, p. 349). They noted several issues. One was that such teams comprised different specialists performing different kinds of work: "stable teams do not seem to be a sufficient answer to the challenge of coordinating complex arrangements of multiple lines of substantially different services which do not rely on each others' assistance or advice on a daily basis" (p. 349). Relatedly, specialists reported that they sometimes had to "contact a number of different professionals and nonprofessionals to quickly coordinate an ad-hoc collaborative effort in order to resolve a problem situation with a client" (p. 349).

Engeström and collaborators provide an example of a sixteen-day sequence dealing with a psychiatric patient, a sequence that ended with the patient being taken into custody. This sequence (based on a single interview with a doctor and an examination of the doctor's records) involved several professionals and nonprofessionals: the patient's general practitioner (GP), a custodian, a psychiatrist, a neighbor, the service technician for the building, a homecare nurse, the police, and an ambulance crew. As the authors emphasize, "This is clearly not an activity system in the sense of having a relatively stable object, motive, community, and division of labor," but "the knot [i.e., the collaborative effort to deal with the psychiatric patient] is not just a singular action either" (p. 352). Nor did a single person or role coordinate all of the actions (p. 352). The authors conclude, "Knotworking is not reducible to a single knot, or a single episode. It is a temporal trajectory of successive task-oriented combinations of people and artifacts" (p. 352). In knotworking, the knot itself is characterized as functioning as "a self-conscious agent": "the subject is the pulsating knot itself, or in other words, subjectivity is dynamically distributed within the knot" (p. 352).

Whereas in Vygotsky's CHAT and Leontiev's CHAT the subject was individual, and in the Westernized CHAT of 1987 the subject was collective, Engeström and colleagues were now considering some subjects to be *distributed*.

The concept of knotworking described fluid collaboration across roles and institutions: "knotworking is dependent on fast accomplishment of intersubjective understanding and distributed control" (p. 354), an emergent form of collaboration that "takes shape without strong predetermined rules or central authority" (Engeström, 2006, p. 1786). In knotworking, different activity systems participate along at least three dimensions: the socio-spatial, temporal, and ethical (pp. 354–355).

Based on the above description, Engeström et al. asked a question demanded by an interventionist approach—and perhaps of interest to their institutional funding sources: "Can knotworking be institutionalized?" (Engeström et al., 1999, p. 356). Their attempt to do so is the *care agreement*, a document that spells out "a plan for the patient's care and the division of labor between the different care providers contributing to the care of the child" (p. 361). The care agreement model "is used to inform practitioners of the new practice" (p. 361). It "aims at resolving the contradictions . . . by creating a new instrumentality" (p. 361), one that represents a durable compromise. In the remainder of the article, the authors describe how this care agreement was deployed, noting two kinds of ruptures:

- Ruptures that indicate that the care agreement should be further refined
- Ruptures that indicate a contradiction between the care agreement's knotworking focus and the hospital's historical division of labor (p. 371).

They conclude by calling a methodological expansion that would set the course for this translation: broadening the scope of the investigation, leading to the spatial and temporal expansion of the object (Engeström et al., 2003).

Here, Engeström et al. undertake a CHAT that *looks* like the design approach from chapter 6: They

- focus on "dialogue, multiple perspectives, and networks of interacting activity systems" (Engeström, 1996b, p. 133);
- take an interventionist stance;
- facilitate the codesign of mediators—in this case, the care agreement, which could mediate knotworking among medical professionals.

Yet differences surface as well, related to the cases they investigated. In the case of the psychiatric patient, the authors noted characteristics different from earlier work in more stable activities:

- Unstable sets of collaborators
- Ad hoc collaboration efforts
- Cases that are not activity systems, but rather draw from multiple allied activity systems (i.e., networks of activity)
- Knots as subjects—not an individual or collective subject, but a temporary collaborative effort (Engeström et al., 1999, p. 352)

The authors connected this case with new work practices. After this case was published in 1999, Engeström and collaborators would go on to develop further conceptual tools to address such cases: knotworking, co-configuration, mycorrhizae, wildfire activities, and runaway objects. These concepts emerged from Engeström's dialogues with other fields. They are distinctive to Engeström's CHAT and reflect its translation to a de facto organizational sociology: an approach that could address larger units and longer timeframes, expanding its scope spatially, temporally, and organizationally.

HISTORY

As we've seen, CHAT has been translated from psychology (chapters 4–5) to interventionist design (chapter 6) to a de facto organizational sociology (this chapter). And along with the movement, these translations addressed different phenomena, bounded in different ways and represented via different kinds of data.

Vygotsky's CHAT and Leontiev's CHAT primarily studied dyads with clear, stable social relations, interacting in closely bounded scenes. Westernized CHAT was applied to somewhat broader cases but still in scenes that were closely bounded in time, space, and participants. These studies typically involved data from specific interactions among specific people with relatively well-defined roles, interacting in a specific place. This boundedness was arguably necessary for the interventionist design of artifacts (such as Bødker's interfaces) and work practices (such as Engeström's doctor-patient consultations and law court cases).

But since Engeström's studies focused on work organizations and sought to intervene in those organizations, producing broad impacts, his cases inevitably broadened in scope. Engeström's studies were not primarily focused on fundamental psychological principles (as Vygotsky's were), nor on pedagogical and rehabilitative applications (as Leontiev's were). Instead, they sought to intervene in specific work organizations with unique configurations and needs. As he discussed in a retrospective interview, "Workplaces and organizations are cradles that generate new forces of production that eventually change the world. Societal contradictions are intensively played out in these settings. Thus, workplaces and organizations are optimal laboratories for observing and analyzing collective and individual development, learning and cognition 'in the wild'" (Engeström & Glăveanu, 2012; cf. Yamazumi, 2020)

To properly investigate these varied work organizations, researchers had to better understand the broader system of each organization. And they undertook this work in the 1990s, when work was rapidly changing due to new information and communication technologies as well as changing political and economic relationships. Thus, studies based in Engeström's CHAT rapidly began to broaden the boundaries of their cases in terms of time, space, participants, and organizations, drawing from the methods and methodologies of social sciences such as sociology and anthropology and yielding more and different data.

By 1995, Engeström and collaborators were investigating activities with broader scope as well as different types of work activities (e.g., Engeström et al., 1995). As they expanded their research scope, examining instances that lasted longer than hours, spanning different settings and multiple data sources, they examined more activity systems connected in broader networks of activity, inevitably leading to more complexity. That complexity involved

- more stakeholders;
- less stability;
- less agreement on what an object is.

But it also interacted with an additional phenomenon, the historical development of work organization.

MORE STAKEHOLDERS

The expanded scope meant including more stakeholders. In previous small-scope investigations, such as investigations of doctor-patient consultations, a small number of actors interacted for a short amount of time as they attempted to address a short-term, defined task. But these small-scope interactions were connected to larger-scope ones, both within and across organizations, as highlighted by the case of the psychiatric patient discussed earlier. In expanding the scope temporally and spatially to understand these larger-scope interactions, Engeström and colleagues necessarily had to expand the set of stakeholders—and thus the number of activities—under examination.

For instance, in the case of the psychiatric patient, stakeholders included the general physician, custodian, psychiatrist, neighbor, service technician, nurse, police, and ambulance crew—far more stakeholders than the dyads and small groups investigated in the previous iteration. Implied in this expansion of stakeholders to organizational and interorganizational scope is a corresponding expansion of contexts and perspectives. Cases become *polycontextual* (Engeström et al., 1995) and *multiperspectival* (Holland & Reeves, 1996), involving *boundary crossing* among historically separate activities (Tuomi-Gröhn & Engeström, 2003). In expanding the number of stakeholders, the researchers also expanded the visible contacts and conflicts among them. That is, investigators had to theorize boundary crossing, in which participants from historically separate activities must interact in concert, and *horizontal learning*, in which these participants must learn enough about each other's activities to work together effectively. Boundary crossing and horizontal learning became increasingly important phenomena not just because of expanded scope but also because cross-functional teams became more broadly prevalent in the late 1990s (as we discuss below).

This issue of multiple stakeholders with differing perspectives is perhaps basic to sociology, but sociology had been banned in the Soviet Union as a bourgeois pseudoscience from the mid-1930s to 1956, during activity theory's formative period (Osipov, 2009; Weinberg 1974), replaced by historical materialism (cf. Sawchuk & Stetsenko, 2008). (Yes, Leontiev believed that one of Soviet psychology's two major tasks was "to define the structure of [human] activity" (Rahmani, 1973, p. 47)—arguably a sociological task—but he did not engage in the empirical work that would necessitate proper concept development.) Thus, Engeström's CHAT had to develop a conceptual and methodological apparatus to address stakeholders more robustly.

LESS STABILITY

Related to the issue of multiple organizational and interorganizational stakeholders was the issue of less stability. Although CHAT had provided an overarching sociocultural framework for conceptualizing durable interactions, empirical work in CHAT had focused on dyads (Vygotsky and Leontiev) and small groups in interactions of very limited scope (early Bødker and Engeström). Expanding the scope temporally and spatially meant acknowledging that different stakeholders could be involved at different points in the timeline and could intermittently bring in different perspectives, understandings, and connections to activities. A closely bounded doctor's visit or court sidebar could include regular disruptions and repairs, but these disruptions and repairs occurred against a backdrop of stable institutional expectations. There was a "script," even if participants could be said to be "twisting" it (Engeström, 1992). In contrast, an expanded-scope case, such as that of the psychiatric patient, involved bringing in different institutions as well as noninstitutional actors at different points. These actors (or stakeholders) may not share scripts, expectations, or conceptual models and may not even be directly coordinated at all.

LESS AGREEMENT ON WHAT AN OBJECT IS

The object of labor activity is both objective (material) and projective (an envisioned outcome), and the activity system forms around the object's cyclical transformation. For a tightly bounded activity, the object can be defined narrowly:

- A hunter-gatherer's object is the quarry (Engeström 1987/2014, chapter 2).
- A blacksmith's object is iron (Engeström, 1990, p. 107).
- A farmer's object is the field that is "transformed time and again from brute earth to crops of grain" (Engeström & Escalante, 1996, p. 360).

In these agricultural and craft examples, activities are strongly and closely bounded by their material objects: we can touch and examine these objects, and we can witness them being transformed in cyclical activity.

But in spatially and temporally expanded cases, different stakeholders from different activities may share the same object but relate to the object from different perspectives as they enter and exit the activity at different points. Furthermore, in such cases, objects also tend to expand, becoming less concrete and more conceptual. For instance, in medical work, there are three potential objects working at different temporal and spatial scales:

- A **diagnosis**, which is typically conducted between a doctor and patient, sometimes with the help of other staff
- **Chronic illness**, which involves "a number of physicians, each viewing the patient from the perspective of his or her own specialty" (Engeström, 2018, p. 25)
- A **treatment plan**, which can involve not just different specialists but also different institutions (hospitals, clinics, support networks).

Larger objects typically involve a longer trajectory including multiple encounters with multiple medical entities working interorganizationally, expanding the unit of analysis temporally and socio-spatially (Engeström et al., 1999; cf. Engeström et al., 1995).

BEYOND SCOPE QUESTIONS: THE HISTORICAL DEVELOPMENT OF WORK ORGANIZATION

Beyond the temporal and spatial expansion of the object, Engeström and colleagues asserted that work has developed historically in different types. For instance, Engeström (1987/2014) characterized activities in terms of craft activity, rationalized activity, humanized activity, and collectively and expansively mastered activity. Later, he (2007, 2008, 2016) drew on Victor and Boynton (1998) to describe historically different types of production (craft, mass production, lean production, mass customization, and co-configuration), each with its own objects and contradictions. In these publications, Engeström argues that these historically different types of production imply different configurations and connections (see Spinuzzi, 2015b for a similar argument about types of work organization). Engeström's CHAT began to be applied more frequently to *knowledge work*: work that is primarily analytic-symbolic in nature.

Here, we discuss these key aspects of CHAT as an organizational sociology: empirical focus, mastery, and mediators.

EMPIRICAL FOCUS: ORGANIZATIONAL AND CROSS-ORGANIZATIONAL COLLABORATION

In its translation into an organizational sociology, Engeström's CHAT broadened its empirical focus, meaning that a new (sociological) phenomenon became visible. Although the unit of analysis continued to be the activity system, Engeström's turn from Vygotsky's and Leontiev's individual subjects to collective subjects (Kaptelinin & Nardi, 2006) had introduced a difference in this unit of analysis, requiring a spatial, temporal, and social expansion of the object and thus a change in the unit of analysis that it anchored. In Westernized CHAT, this change meant larger and larger scenes. Engeström and colleagues began to draw from qualitative research approaches such as ethnography and conversation analysis. As it was translated from a psychology to a design approach to an organizational sociology, Engeström's CHAT required a *sociological* unit of analysis and *sociological* methods.

But like the design approach of Westernized CHAT, this sociological work was *interventionist*. Whereas Leontiev had *acknowledged* contradictions as besetting capitalist activities, Engeström, following Ilyenkov, made contradictions *central* to the account of all change in activities. These contradictions were both internal to a given activity system (embedded in the activity, its components, and the relationships among them) and external to it (across a network of activity systems). For Engeström, these contradictions were the engines of change: the results manifesting from these contradictions engendered interventions from the actors in the activity. So, through his interventionist approach, Engeström attempted not just to study activities but to coordinate interventions led by these actors. Interventionism required a unit of analysis that allowed both researchers and participants to *explore, describe, and deliberate over* social relations and contradictions—a sociological task—rather than *assuming* them as Leontiev's historical materialist activity theory had done.

The newly sociological scope of empirical focus meant that sociological phenomena became newly visible and had to be addressed through new conceptual tools. But complicating the picture was the fact that work organization itself was also changing during this time (Spinuzzi, 2015a). Thus, it is sometimes difficult to sort out which conceptual innovations in this period are due to the new scope versus which ones describe newly prevalent aspects of work organization.

The scope of this sociological task broadened as CHAT was applied in different countries, with different cultural and economic assumptions, as well as to different types of work.

THE CHANGE LABORATORY METHODOLOGY

The change laboratory methodology, which was based on the cycle of expansive learning (see chapter 6), was introduced in the mid-1990s (Engeström, 2011). It was focused on developing "formative interventions based on Vygotsky's principle of double stimulation" (Engeström, 2011, p. 598). "In formative interventions, the researcher aims at provoking and sustaining an expansive transformation process led and owned by the practitioners" (p. 606). In contrast to Engeström's earlier developmental work research, it was not restricted to the conversation analysis of restricted vignettes. In contrast to Bødker's earlier work, it provided an approach that was more retrospective and group-oriented than Bødker's interventionist methods (mockups, prototyping, organizational games; cf. Spinuzzi, 2005b), while being more structured and formally data-driven than the future workshops that her participatory design colleagues had adopted from Jungk (Jungk & Mullert, 1988; Ehn, 1989).

Engeström described the change laboratory methodology in a 2011 retrospective:

> A Change Laboratory is typically conducted in an activity system that is facing a major transformation. This is often a relatively independent pilot unit in a large organization. Working practitioners and managers of the unit, together with a small group of interventionist-researchers, conduct five to ten successive Change Laboratory sessions, often with follow-up sessions after some months. When feasible, customers or patients are invited to join Change Laboratory sessions, in which their particular cases are analyzed in detail. Change Laboratories are also conducted as boundary-crossing laboratories with representatives from two or more activity systems engaged in collaboration or partnership. (2011, p. 612)

(Engeström does not specify how he convinced people to participate in these change laboratories, but many change laboratory studies list institutional funding sources, as seen at the beginning of this chapter.)

Change laboratories have been conducted in organizational settings including post offices, factories, schools, hospitals, newsrooms (Sannino & Engeström, 2017), eldercare programs (Nummijoki & Engeström, 2013), libraries (Engeström et al., 2012; Sannino et al., 2016), and organic agriculture programs (Mukute et al., 2018).

During this period, Engeström and colleagues undertook a great deal of theoretical development to better address the phenomena they discovered as they broadened their empirical cases beyond the dyad. As we've seen, these phenomena are often attributed to historically new types of production (e.g., Engeström, 2008, pp. 16–17). Yet they arguably *also* reflect the spatial, temporal, and social expansion of the cases under study. That is, they had already existed, but were newly visible to CHAT.

To understand historically different types of production, activity theorists needed to understand and explore broader networks of activity. By the mid-1990s, Engeström and collaborators began noting that such cases did not fit neatly into well-bounded activities, involving unstable sets of interlocutors, ad hoc coordination, and multiple specializations. They thus developed further conceptual tools to address such cases—conceptual tools that are distinct to Engeström's CHAT. These concepts, which included *co-configuration, knotworking, mycorrhizae,* and *runaway object* (table 8.1), did not have parallels in earlier CHAT work, and many were either borrowed from existing managerial and social theory (e.g., co-configuration) or developed in contradistinction to existing concepts from those disciplines (e.g., mycorrhizae and runaway objects). Reflecting this intense interdisciplinary borrowing and dialogue, Engeström's CHAT descriptions in this period sometimes involved wildly mixed metaphors—metaphors that were directed outward, attempting to perform interessement across different conversations in managerial and social theory.

These concepts translated Engeström's CHAT further, enabling it to function as an organizational sociology.

CO-CONFIGURATION

Engeström and colleagues define *co-configuration* as "an emerging, historically new type of work," one that "requires flexible 'knotworking' in which no single actor has the sole, fixed authority—the center does not hold" (Engeström et al., 1999, p. 24). Co-configuration is set within a larger sweep

Table 8.1

Concepts, definitions, and sources for CHAT as organizational sociology

Concept	Definition	Source or contradistinction
Co-configuration (Engeström, 2007)	"An emerging, historically new type of work that relies on (a) adaptive 'customer intelligent' product service combinations; (b) continuous relationships of mutual exchange . . . (c) ongoing configuration and customization . . . (d) active customer involvement and input into the configuration; (e) multiple collaborating producers . . . and (f) mutual learning from interactions. . . ." (Engeström, 2007, p. 24)	Co-configuration (Victor & Boynton, 1998, p. 195).
Knotworking (Engeström et al., 1999)	"Knotworking is characterized by a pulsating movement of tying, untying and retying together otherwise separate threads of activity. The tying and dissolution of a knot of collaborative work is not reducible to any specific individual or fixed organizational entity as the center of control." (Engeström et al, 1999, p. 346)	Networks (Barley, 1988)
Mycorrhizae (Engeström, 2006)	"Communities are increasingly beginning to take the shape of weakly bounded and heterogenous mycorrhizae-like formations which rely on stigmergy, negotiation and peer review as coordinating mechanisms" (Engeström et al., 2010, p. 8)	Rhizomes (Deleuze & Guattari, 1987)
Wildfire activities (Engeström, 2009b)	"Wildfire activities pop up in unexpected locations at unexpected times and expand very rapidly. They become extinguished from time to time, yet they reappear and flare up again. Learning in wildfire activities is learning by swarming that crosses boundaries and ties knots between actors. It is also learning by building mycorrhizae communities by means of cognitive trails and social bonds that make the terrains knowable and livable. The mechanism of stigmergy is foundational in mycorrhizae communities." (Engeström, 2009b, p. 1)	None
Runaway objects (Engeström, 2008)	"Runaway objects have the potential to escalate and expand up to a global scale of influence. They are objects that are poorly under anybody's control and have far-reaching, unexpected side effects." (Engeström, 2008, p. 227)	Runaway world (based on Giddens)

of historically and materially developing work organization. The roots of this account are in *Learning by Expanding*, chapter 4 (Engeström, 1987/2014), which discusses the historical development of work activity. But in drawing on Victor and Boynton (1998), Engeström names successive types of work organization, with the most recent being co-configuration (cf. Castells, 2003). In this form of work, the activity system is no longer sufficient: "The emerging third generation of activity theory takes two interacting activity systems as its minimal unit of analysis" (Engeström, 2001, p. 143). An activity network forms a tight interconnection across interacting activity systems, leading Engeström's CHAT to focus on quaternary contradictions (contradictions across these interacting activity systems).

But how is co-configuration organized? Engeström's answer was knotworking.

KNOTWORKING

Engeström and colleagues define a *knot* as "a rapidly pulsating, distributed and partially improvised orchestration of collaborative performance between otherwise loosely connected actors and activity systems" (Engeström et al., 1999, p. 347).

Knotworking is defined as "a historically significant new form of organizing and performing work activity, connected to the emergence of new co-configuration models of production." In it, "the locus of initiative changes from moment to moment within a knotworking sequence. Thus, knotworking cannot be adequately analyzed from the point of view of an assumed center of coordination or control, or as an additive sum of the separate perspectives of individuals or institutions contributing to it. The unstable knot itself needs to be made the focus of analysis" (pp. 347–348).

Here and elsewhere (Engeström, 2006, p. 1786; Engeström, 2008), knotworking is characterized as a new form of work, one that enables the synthesis of different perspectives and activities seen in co-configuration work. Yet we may have doubts about how new knotworking is. Engeström et al. (1999) proffer knotworking to explain the extended intermeshing of work across emergency services, routine medical services, and other actors. But we see such intermeshing in older sociological studies of society (e.g., Whyte, 2012) and work (e.g., Whyte & Hamilton, 1964). The phenomenon was not historically new, but it had become *newly visible* to Engeström's CHAT.

Knotworking was applied to at least four sorts of cases, raising questions about how precise a concept it is (see table 8.2).

These four types of cases are different in:

- collaborators;
- relationships among collaborators;
- communication, coordination, and collaboration mechanisms.

Yet they were collectively addressed under the heading of knotworking, as collaborations that do not fit the initial (intrainstitutional or dyadic) expectations of previous iterations of CHAT.

Knotworking is founded in community, but a specific type of community: mycorrhizae.

MYCORRHIZAE

Engeström defines *mycorrhizae* as a social formation, one that "typically does not have strictly defined criteria of membership. But its members can be identified by their activism. . . . A mycorrhizae formation is simultaneously a living, expanding process (or bundle of developing connections) and a relatively durable, stabilized structure; both a mental landscape or 'mindscape' . . . and a material infrastructure" (Engeström, 2006, p. 1788). He adds that "horizontal and invisible mycorrhizae do not eliminate visible, erect, bounded and institutionalized activity systems"; rather, "relatively stable and well-bounded" activities are a necessary precondition for their formation (Engeström, 2006, p. 1788). In these mycorrhizal communities, otherwise separate activities are interlaced through horizontal, invisible links. He offers the example of the Linux community, in which

> the 2,605 "development team" members and 1,562 "bug reporting team" members . . . were identified on the basis of their publicly available contributions to the development and perfection of the object, the Linux operating system. It is very likely that mycorrhizae include quite a variety of members, ranging from grass-roots activists, or clients or victims, to certified professionals, researchers, enterpreneurs [sic], and spokespersons. (Engeström, 2006, p. 1788)

The members of this development community are volunteers that represent different kinds of expertise and motivation, and they are not strongly coordinated. Still, this community *emerges* from more stable activities: software developers and researchers, for instance, belong to professions,

Table 8.2
Four applications of knotworking

Application	Description	Examples
Institutions temporary mesh around a shared object	An institution is a lead activity, containing different specialties that share an object (e.g., patient's health). Learning occurs horizontally across disciplinary boundaries. Individuals outside the institution may participate intermittently. Leadership rotates as different expertise is required.	Engeström et al., 1999; Engeström, 2008
An institution engages outside (noninstitutional) expertise	An institution asks for external expertise (consulting, contracting, other services). Collaboration is part of the service and is project-oriented, with the institution initiating the project and defining success conditions. However, external expertise defines the details of an emergent project.	Engeström et al., 2012; Kaatrakoski & Lahikainen, 2016; Kerosuo, 2018; Korpela, 2015
An institution internally leverages cross-functional teams	An institution develops cross-functional teams. The lead institution underwrites objects and rules and initiates and defines a project. People positioned within the institution's hierarchy are loaned to a cross-functional team to work on projects of bounded scope and duration.	Daniels et al., 2007; Kerosuo, 2015; Kerosuo, 2017; Kerosuo & Toiviainen, 2011; Korpela & Kerosuo, 2014
Noninstitutional actors form noninstitutional, nonproject collaborations	These cases typically do not include a lead institution. They are not new (Engeström, 2009b) but are arguably supported at a larger scale by digital media. They are nonhierarchical, with voluntary and intermittent engagement. They involve localized projects and unpredictable pulses. Mediators are typically borrowed from elsewhere.	Engeström, 2008, 2009b, 2016, 2018

are educated in specific bounded ways, and work for stable, well-bounded organizations.

Engeström positions this notion of mycorrhizae against Deleuze & Guattari's (1987) concept of *rhizome*, which "highlights the importance of horizontal and multidirectional connections in human lives, in contrast to the dominant vertical, tree-like images of hierarchy" (Engeström 2006, p. 1787), but which Engeström finds to be too limiting. In doing so, Engeström implicitly rejects the pure association of rhizomatics, abducting the notion of horizontal connections while retaining the developmental, dialectical orientation inherent in CHAT. As Engeström explains, literal mycorrhizae are "the invisible organic texture underneath visible fungi" (p. 1787): a material though invisible substrate that underlies development and that, undetected, can give the illusion of spontaneous and unrelated growth. Mycorrhizae "are difficult, if not impossible, to bound and close, yet are not indefinite or elusive" (p. 1788), and similarly, he asserts, figurative mycorrhizae underlie the horizontal and seemingly associational connections across historically separate activities. Engeström and collaborators develop this concept across multiple publications (Engeström, 2006, 2008, 2009b, 2016; Engeström et al., 2010).

Mycorrhizae underlie a final concept, *wildfire activities*.

WILDFIRE ACTIVITIES

Wildfire activities "have the peculiar capacity to disappear or die in a given location and suddenly reappear and develop vigorously in a quite different location, or in the same location after a lengthy dormant period" (Engeström, 2006, p. 1785). These activities may surface, thrive, then "burn out," only to "flame up" again. Engeström and Sannino later characterized wildfire activities through analogy to the zone of proximal development attributed to Vygotsky:

> In wildfire activities, a zone is a terrain to be dwelled in and explored, not just a stage to be achieved or a space to be crossed. The zone is explored by moving in it, to various directions and destinations, back and forth and sideways. The dwellers create trails and the intersecting trails gradually lead to an increased capability to move in the zone effectively, independently of the particular location or destination. The zone is never an empty space to begin with; it has dominant trails and boundaries made by others, often with heavy histories and power invested in them. When new dwellers enter the zone, they eventually have critical encounters with existing trails. They both adapt to the dominant trails and struggle to go beyond them. The latter can lead to new trails that expand the collective shape and

> understanding of the zone, thus also to new boundaries. When the dwellers reach a certain level of mastery of the zone, they begin to collide with the very boundaries of the zone and to break away from the zone, toward new zones. (Engeström & Sannino, 2010, p. 21)

Frankly, this is a very difficult passage for readers to disentangle due to the many concepts, expressed with wildly mixed metaphors—wildfires, zones, dwellers, trails, encounters, and collisions—as well as the application of the zone of proximal development to indistinct groups. Adding yet another metaphor, Engeström describes wildfire activities as "amoeba-like collective activities that are not limited to the pursuit of short-term goals" (Engeström, 2006, p. 1785). But this cacophony of metaphors reflects the different communities that Engeström was trying to interest in his CHAT.

To illustrate his points, Engeström offers examples such as birding and skateboarding. Both are

- domains, but not activity systems in the sense that has been used up to this point in CHAT;
- *aided* by new information and communication technologies, although they *predate* those technologies (Engeström, 2006, p. 1787) (that is, they are not historically new forms of collective action);
- "peculiar combinations of leisure, work, sport, and art. And they both have consistently defied attempts at full commercialization, offering ample opportunities for entrepreneurship but not becoming themselves dominated by commercial motives" (Engeström, 2006, p. 1785).

And both interest Engeström because "large portions of [the] population find it increasingly difficult to find meaning and motivation in the financialized spheres of work. There is breeding ground for alternatives" (Engeström, 2006, p. 1786)—that is, they may point to alternatives to capitalism, as we'll see in chapter 9.

These four conceptual developments—co-configuration, knotworking, mycorrhizae, and wildfire activities—have a common theme: what *appears* spontaneous, temporary, and unstructured is *analyzed as* actually being more durable and structured. That is, what appears associational is interpreted as dialectical—an assumption that Vygotsky also makes at the individual level (Vygotsky, 1934/2012, chapter 5). But this is an assumption, an article of faith, not an empirically verified proposition.

MASTERY: OF ORGANIZATIONAL AGREEMENT

As we've seen, mastery has been translated considerably:

- In Vygotsky's CHAT, a person mastered themselves.
- In Leontiev's CHAT, a person accepted state mastery.
- In Westernized CHAT circa 1987, mastery was exerted by collective participants in an interventionist study, conducting a dialogue in order to develop mediators.

Each translation involved a progressive widening of the scope of empirical investigation. And in Engeström's CHAT, that scope had widened enough that mastery could *only* be collective and partial: it was achieved collectively during interventionist studies. The wider scope meant that the anchor of an activity, the object, was not apprehendable in its totality by any single actor.

Such objects are characterized using yet another metaphor defined in contradistinction to existing social theory. *Runaway objects* "have the potential to escalate and expand up to a global scale of influence. They are objects that are poorly under anyone's control and have far-reaching, unexpected side effects" (Engeström, 2008, p. 227). Examples include global warming, narcotics, Linux, and chronic illness. Runaway objects start as small problems or marginal innovations but balloon into objects that are larger than any of the activity systems that are oriented toward (and defined by) them (p. 227). Such objects are

- no longer closely bounded or material;
- much bigger than the materials in which they are instantiated;
- multiperspectival;
- the nexus of many different activities.

These objects are characterized as historically new and mastered only in part (Engeström, 2008). But as discussed earlier, we might question whether these are actually historically new: objects that "are poorly under anyone's control and have far-reaching, unexpected side effects" could include the economy, democracy, or world peace. Such objects surely predate any version of CHAT. In any case, Engeström bases this formulation on Giddens' "runaway world" (Engeström 2006, p. 1784).

Although a runaway object is too big for any individual to understand, Engeström still understands it as unitary. It might be *fractured*—that is, the object is multiperspectival, understood differently from different perspectives—but is ultimately a unified, ontologically stable and shared phenomenon.

Given the discussion above, researchers must take an interventionist stance in which participants collectively develop and articulate mastery. Through this interventionist work (especially through the change laboratory methodology), participants engage in horizontal learning and boundary crossing, yielding a new, collectively developed concept of work (e.g., Sannino & Engeström, 2017).

MEDIATORS: SECOND STIMULI AND COORDINATION GENRES

The scope in previous translations of CHAT was small enough that researchers could focus on singular mediators (Vygotsky's signs, Leontiev's colored cards) or small clusters of mediators (Bødker's interface elements, Engeström's patient records). But as Engeström and colleagues developed the change laboratory methodology and related interventionist approaches, the scope expanded to the organizational or even the extra-organizational level, meaning that an exhaustive census of mediators in the activity would be unproductive, if not impossible. Naturally, then, the focus of mediation in this iteration shifted to two classes of mediators—classes that continue the split between designing and designed mediators we saw in chapter 6.

SECOND STIMULI: DESIGNING MEDIATORS

The first class of mediators is "second stimuli." In a change lab, these designing mediators come in two classes:

- **Mirror materials** are ethnographic data, such as interviews and photographs. Participants and an interventionist researcher examine these materials "to stimulate collective analysis, problem-solving and developmental efforts by the participants" (Haapasaari et al., 2016, p. 237).
- **Conceptual tools** are then used to collectively interpret and analyze the mirror materials. These tools include the activity system triangle and historical matrices (cf. Engeström, 2011, p. 612). The conceptual tools are used "as an instrument in the design of a new concept for the activity [participants] are trying to transform" (Engeström, 2011, p. 613).

Together, mirror materials and conceptual tools provide a double stimulation in which participants contemplate both the problem and its re-representation (Sannino, 2011, built on the double stimulation technique described in Vygotsky, 1978). That is, they are used to *conduct* change labs.

COORDINATION GENRES: DESIGNED MEDIATORS

The second class of mediators—the *output* of change labs—is that of *coordination genres* such as agreements and organization charts. By "genre," I mean "typified rhetorical actions based in recurrent situations" (Miller, 1984, p. 159; see chapter 7). That is, a genre is a patterned rhetorical performance that has historically worked well to address a recurrent rhetorical situation. For instance, Bazerman (1988) traces the historical evolution of the experimental article genre, noting how it changed over time to address recurrent concerns: by imitating previous, successful experimental articles, scientists were able to avoid rhetorical missteps by anticipating and preemptively addressing recurrent questions. A genre is not a template so much as an orientation to the recurrent situations—and, more broadly, the activities—in which it has developed (Russell, 1997a; Spinuzzi, 2003b).

Critically, as CHAT was translated from an interventionist design approach to an organizational sociology, the design focus changed from individual artifacts (such as an interface) or practices to genres that could function as *charter documents*: "The charter document of a social or political group establishes an organizing framework that specifies what is significant and draws people's attention to certain rules and relationships. In other words, the charter defines as authoritative certain ways of seeing and deflects attention from other ways. It thus stabilizes a particular reality and sets the terms for future discussions" (McCarthy, 1991, p. 359).

One example of a charter document is the *Diagnostic and Statistical Manual of Mental Disorders* (*DSM*), which establishes and legitimizes base categories for understanding mental disorders (McCarthy 1991). Similarly, in a series of studies on medical work, Engeström and collaborators describe the development of the care agreement for health care (Engeström, 2001, 2005, 2018; Engeström et al., 2003, 2006) and the mobility agreement for eldercare (Nummijoki & Engeström, 2013; Nummijoki et al., 2018). In these cases, coordination is the key in meshing efforts: a horizontal agreement across participants that governs the other tools, the relevant rules, and the division of labor in a particular case of knotworking.

Coordination genres are a natural output for Engeström's CHAT-as-organizational-sociology:

- They are appropriate to the newly expanded scope.
- They facilitate the horizontal learning that is required in multidisciplinary activities.
- They can emerge from deliberation, representing a meshing of views from different activities in the network.

That last point makes coordination genres more consequential. At the level of organizational sociology, any given activity is mediated by multiple texts and tools; just imagine the number of tools and texts that are involved in health care or eldercare. Each activity has several, each of which has developed over time, and they have been conceived as interacting to form *compound mediation* (Spinuzzi, 2003b) or *complex mediation* (Bødker & Andersen, 2005). Rather than trying to redesign several of these, Engeström's CHAT-as-organizational-sociology seeks to develop a new coordination genre, a single point of change that can fundamentally reorganize the activity and its many other mediators.

SUMMARY: TRANSLATING CHAT TO INTERVENE IN AND ACROSS ORGANIZATIONS

The first halting steps toward Engeström's CHAT were taken in the 1980s (chapter 6). By the 1990s, Engeström and colleagues had begun constructing a new set of concepts to translate Engeström's CHAT into an organizational sociology, in contradistinction with existing concepts and studies in that field. They refitted Vygotsky's concepts (the zone of proximal development, double stimulation) and Leontiev's concepts (community, contradictions) to produce an approach that could both model organizations and intervene dialogically in them. As the Soviet Union crumbled, this translation reoriented Engeström's CHAT to Western concerns in the late twentieth century: how organizations worked and how they would change as knowledge work became ascendant. Engeström's CHAT gained a modeling approach for organizations as well as the spatial, temporal, and social expansion of the object required for studying phenomena at a larger empirical scope.

But in taking up CHAT, Engeström and collaborators translated it in key ways:

- It added still more stakeholders, representing more activities involving larger numbers of people.

- It used social and managerial theory to explore loss of stability in these larger activities and networks of activity.
- It reflected disagreements over what object a given activity had.
- It reflected a historical narrative about a shift toward networked work organization.

These changes can be summed up in three different translations:

- **Empirical focus**: CHAT's empirical focus widened from relatively small groups to organizations and interorganizational collaborations.
- **Mastery**: Its understanding of mastery shifted from codesigning mediators to generating an organizational agreement.
- **Mediators**: As the empirical focus widened, bringing in far more mediators, its mediational focus shifted to coordinative mediators that could reflect and lock in that organizational agreement. To achieve those designed mediators, change labs incorporated ethnographic and conceptual tools (second stimuli) to use in deliberations.

Through these translations, Engeström and colleagues

- **problematized** managerial and social theory, specifically in terms of interorganizational dynamics;
- **interested** a broader number of fields and disciplines, making their arguments to sociologists, management scholars, vocational education researchers, and many others, as well as institutional funders and participants themselves in various institutions;
- **enrolled** managerial and social theorists such as Barley, Deleuze and Guattari, Giddens, and Victor and Boynton, as well as institutional funders and the institutions that served as their research sites;
- **mobilized** all of these enrolled actors, using change laboratories to yield organizational transformation.

The resulting organizational sociology (table 8.3) could generate agreed-upon solutions. Like the last translation, this one turned out to be quite successful, and Engeström's CHAT was picked up across the globe in an increasing number of fields and disciplines.

HOW WELL DID ENGESTRÖM'S CHAT WEATHER ITS TRIALS?

As we've seen, this iteration of CHAT involved a rapid expansion of research scope and corresponding conceptual tools and framework. This expansion

Table 8.3
Domain, project, translations, and trials in Engeström's CHAT-as-organizational-sociology

Domain	Organizational sociology
Project	Reorganization of interorganizational work
Translations	
• ***Empirical focus***	Organizations and interorganizational collaborations
• ***Mastery***	Generating an organizational agreement
• ***Mediators***	Charter documents to reflect organizational agreement
Trials	Stable sets of interlocutors; partial integration with managerial and social theory; boundaries of activity; instrumentality

led Engeström's CHAT to intervene more broadly, addressing phenomena that have typically been addressed in sociology. The expansion also happened against a background of rapid work changes.

The translation of Engeström's CHAT into an organizational sociology continued a trial and introduced three new trials:

- **Stable sets of interlocutors**: The continuing trial, left over from CHAT's iteration as a design approach, was that the widening scope meant further instability in sets of interlocutors. A wider scope meant more *potential* participants, not just in the activity under scrutiny (in change laboratories; e.g., Engeström, 1999a) but also in impinging activities (in boundary-crossing laboratories; e.g., Engeström 2001). Yet *actual* participants were limited to a relatively small number of representatives; these were qualitative studies, after all. And, critically, participants were generally functional representatives, chosen for their job roles, rather than political representatives authorized to speak on behalf of unions. In one example, "the participants of the sessions were selected to represent the whole range of practitioners working in the unit, from the head doctor of the unit to surgeons, anesthesiologists, nurses, a porter, and a secretary" (Engeström, 2001, p. 613). Yes, change laboratories can be chained to work with rotating sets of stakeholders, but a change laboratory "is often a relatively independent pilot unit in a large organization" (p. 612).
- **Partial integration with managerial and social theory**: One new trial was that although the translation yielded an organizational sociology by incorporating existing concepts into the Engeström's CHAT matrix, this integration was partial, yielding an uneasy, unstable settlement characterized

by mixed metaphors and a ballooning conceptual apparatus. Terms were lifted from elsewhere rather than organically developed, oriented to extending or rebutting existing conversations in the space rather than focusing on internal coherence. CHAT was already complicated by the time Leontiev was finished with it—but now, it was rapidly incorporating many new terms and concepts that required understanding the interdisciplinary conversations in which Engeström was enmeshed (see especially table 8.1). From a sociology-of-translation perspective, this complication creates severe hurdles for those who seek to take up Engeström's CHAT and translate it for their own needs.

- **Boundaries of activity**: Relatedly, another new trial was that of the boundaries of activity. The greatly expanded empirical focus made broader organizational phenomena visible, some of which were well-known issues in organizational sociology, others of which represented new phenomena related to changes in work organization in the 1990s and early 2000s. In examining such phenomena, Engeström and collaborators found it increasingly difficult to identify and articulate the boundaries of any given activity. That is, the conceptual tools of Engeström's CHAT had begun to show severe limitations when applied to problems that went beyond well-defined organizations.
- **Instrumentality**: Another new trial was that of the instrumentality of Engeström's CHAT. As we'll discuss in chapter 9, throughout the early 2000s, Engeström's CHAT came under criticism from others working in the CHAT tradition for its perceived instrumentality and apoliticism. These criticisms focused on what was perceived to be a tight, instrumentalist focus on improving inter- and intraorganizational functions. Although Engeström had declared that any capitalist society faced the primary contradiction between exchange and use value (1987/2014, p. 102), critics argued that this primary contradiction was not actually addressed in his work and that capitalism had been insufficiently criticized. (This criticism echoes criticisms that Leontiev faced due to perceived apoliticality; see Maslina, 1950.)

Thus CHAT-as-organizational-sociology represented an unstable settlement, and this instability set the stage for the next translation. The latter two trials were addressed in Engeström and Sannino's work from 2009 on, work that translated Engeström's CHAT further into the domain of public policy.

9 ENGESTRÖM'S "FOURTH-GENERATION" CHAT: AN IMPLEMENTATION OF PUBLIC POLICY

In chapter 3, I mentioned Engeström's generational account of CHAT. Engeström (1996b; 2001) and Sannino (Engeström & Sannino, 2021; Sannino, 2011; Sannino & Engeström, 2018) have argued that activity theory has developed in three distinct generations. In the first generation, Vygotsky developed the unit of analysis of culturally mediated action. In the second, Leontiev developed a second unit of analysis, activity. In the third, Engeström and collaborators developed a third unit of analysis, the activity network, which consists of minimally two activity systems sharing the same object.[1]

In a response to chapters in his 2009 Festschrift (Sannino et al., 2009), Engeström recognized that this third unit of analysis may be inadequate for addressing poorly bounded activities (see chapter 8). Using examples such as open source software and Wikipedia (Engeström, 2009a, pp. 309–310), he suggested that "we are moving toward increasingly open, amoeba-like communities characterized by multidirectional swarming, weak boundaries, and no stable center," communities in which authority and agency grow from minority influencers from the peripheries (2009a, p. 317), even though they may involve "bounded hubs of concentrated coordination efforts" (p. 310). In such cases, Engeström argues, a fourth generation of activity theory may be needed to address "social production and peer production," in which "the boundaries and structures of activity systems *seem* to fade away" (p. 309, my emphasis).

This statement has been interpreted in various ways, with some (including myself) interpreting it as updating CHAT's organizational sociology to address loosely bounded postbureaucratic organizations that had been empowered (and intermeshed) through changes in information and communication technologies (see Spinuzzi & Guile, 2019 for a review). But in later publications of their own as well as interviews with other scholars, Engeström and Sannino (Engeström & Sannino, 2021; Ploettner & Tresseras, 2016; Yamazumi, 2020) made clear that they regarded fourth-generation activity theory as being about alternatives to capitalism (Engeström & Sannino, 2016, p. 403). That is, they were interested in *postmarket society*, not *postbureaucratic work* (see Acquier et al., 2017). Thus, their efforts in this fourth generation addressed learning in social issues and movements, movements that were built on mycorrhizal communities' subterranean connections across existing, stable activities. These efforts consequently expanded empirically far beyond the scope of Vygotsky's and Leontiev's dyads, the small bounded groups of Engeström's early work, or even the organization-specific change laboratories of Engeström's 1990s and 2000s work. The social issues addressed included childbirth support, sustainable food cooperatives, social justice, and homelessness. That is, they entail a shift to a *method for implementing public policy*.

In applying CHAT to public policy, Sannino and Engeström introduced a concept of the public that has heretofore been missing in Engeström's CHAT: that of "heterogeneous coalitions," a new unit of analysis in which multiple groups of people engage in interdependent cycles of expansive learning as they orient to a shared runaway object (Engeström & Sannino, 2021; Sannino, 2020a). Here, the familiar triangles disappear. This approach has been taken up by a growing array of interventionist researchers and applied to public policy objects such as social inequities in education (Bal et al., 2021) and court systems (Lémonie et al., 2021), as well as community-based learning in farming (Lotz-Sisitka et al., 2017; see the end of this chapter for more on this research) and other complex, large-scale social problems (Engeström et al., 2016). Significantly, the focus in this phase moves from organizations to the larger public, from acting-with-technology to policy implementation.

To better understand Engeström's CHAT as an implementation of public policy, let's again examine *domain*, *project*, *translations*, and *trials*.

- **Domain**: The domain that Engeström and Sannino addressed in this period was public policy. They translated insights from Engeström's CHAT-as-organizational-sociology so they could apply its change labs methodology and conceptual apparatus to public policy issues such as homelessness, global warming, and pandemics. Surprisingly, in contrast with the previous period, they did *not* borrow from existing public policy approaches, instead relying primarily on their own previous work on reorganizing interorganizational work.

- **Project**: Engeström and Sannino's project—what they were trying to achieve—was the implementation of public policy objectives to address acute societal problems. This overarching project, which we'll discuss below, drove their adaptation of Engeström's organizational sociology for public policy objectives. It is often described as a "fourth-generation activity theory" (Engeström & Sannino, 2021, p. 13) that is meant to produce "alternatives to capitalism" (p. 4).
- **Translations**: To produce this method for implementing public policy, Engeström and Sannino translated CHAT in a few ways: by developing a larger unit of analysis (heterogeneous coalitions) and by adapting the change labs approach to produce "agency" rather than documented agreements. We'll specifically examine translations in empirical focus, mastery, and mediators. These translations are still ongoing, so it's unclear what settled method for implementing public policy will result.
- **Trials**: Finally, this nascent settlement is currently undergoing trials, including continued skepticism from outside and inside CHAT circles.

To examine how this ongoing translation impacts Engeström's CHAT, let's first look at a paradigmatic case and then examine the history of this case's development and its stance on empirical focus, mastery, and mediators.

AN EXTRAORDINARY CASE: ERADICATING HOMELESSNESS

In the mid-2010s, Sannino began a multiyear change laboratory effort in coordination with the Finnish Housing First strategy, an approach that seeks to eradicate homelessness by "supplying affordable housing and targeted services to a population with complex problems and high need of social and health care" (Sannino, 2018, p. 386). In contrast to traditional public policy addressing homelessness, Housing First "focuses first on attaining and securing housing for the clients, in combination with treatment plans that they individually negotiate with service providers. With the Housing First model one['s] own [housing] is guaranteed even to the clients who fall out from treatments or are not yet ready to engage with the support services" (Sannino, 2018, p. 386).

Finland's Housing First initiative, Sannino tells us, "is a utopia that was realized: Finland is the only country in Europe that has had a consistent reduction in the numbers of homeless people since 2008" (quoted in Yamazumi, 2020, p. 9). This initiative entails education: "counteracting the stigma of homelessness requires re-educating actors and institutions to engage in interactions in which the Housing First client is an interlocutor and a responsible agent" (Sannino, 2018, p. 390). More broadly, how could the world *enact a utopia* relative to homelessness? Sannino, who sees herself as a "hunter of utopias" (p. 9), understands this specific problem of homelessness as a complex problem requiring the coordination of multiple activities at multiple levels: "Long-term homelessness is prolonged living without one's own home due to socio-economic, health- and addiction-related problems. As such it is a web of multiple conditions which is considered a top indicator of poverty associated with extreme social exclusion and vulnerability. A problem like this requires a peculiar organizational form that we call heterogeneous work coalitions" (Sannino & Engeström, 2018b, p. 72).

This "heterogenous work coalition of the Finnish homelessness strategy"—constituted by "individual clients and their frontline support workers to housing units, NGOs, ministries and municipalities"—is understood "as intertwined cycles of expansive organizational learning" (p. 73), that is, "multiple interacting and overlapping cycles which are both relatively independent and interdependent . . . [requiring] one another to emerge and sustain themselves. They also operate across sectors and at different hierarchical levels in the society" (pp. 73–74). In this coalition,

> each client has a specific constellation of issues to address which require supportive interactions and tailored solutions co-designed with the client, and changing as the client progresses or regresses. What is peculiar of these intertwined learning processes is their strong agentive nature. It is work made of tailored concerted efforts by a flexibly developing work coalition pursuing one and the same motive and purpose: the functional independent living of the client in a home of his or her own. (p. 74)

Like the problem of the psychiatric patient discussed in chapter 8, the problem of homelessness is broad in scope, requiring multiagency collaboration. But unlike that earlier case, homelessness is a continuing societal problem requiring a broader, longer-term commitment to durable policy goals. It requires a method for implementing public policy.

To pursue these overlapping cycles of expansive learning, Sannino and colleagues used the change laboratory methodology discussed in chapter 8—but with an interesting change. The mechanisms for these change laboratories include coordinating genres such as lease agreements, support contracts, service plans, and letters of intent. But these "agreements and plans are established and systematically used as *second stimuli*" (Sannino & Engeström, 2018b, pp. 74, my emphasis; cf. Engeström, 2023), and "by means of these documents *the agency of the coalition as a whole is built* by grounding it in specific local commitments" (pp. 74–75, my emphasis). In CHAT-as-organizational-sociology, coordinative genres were *designed*, the product of change labs. In the Housing First case, they were lumped in with second stimuli as designing documents. What was being designed through them was the coalition's agency.

These agreements and plans, used as second stimuli, address a binary dilemma that Sannino identifies as facing individual homeless clients: "The clients are torn between living a healthy life in the apartment rented to them vs. continuing to use drugs and alcohol, for instance. The signed rental agreement and negotiated care plans serve as mediators and supporting agency formation anchors, what Vygotsky called 'second stimuli.'" (Sannino, 2018, p. 389; cf. Sannino, 2020b).

Why had agreements shifted from designed mediators to designing mediators? Because the change labs were no longer design activities. "Counteracting the stigma of homelessness," Sannino argues, "requires *re-educating actors and institutions* to engage in interactions in which the Housing First client is an interlocutor and a responsible agent" (2018, p. 390, my emphasis). Put another way, in this project, the policy objective (eliminating homelessness) and policy (Housing First) are already decided upon and set. The "utopia" has already been settled. *The actors' roles are to be persuaded to actively comply* in two ways: (1) to be educated to engage in interactions, and (2) to identify ways to implement the socially correct choice in the dilemma.

The multitude of problems related to homelessness—from mental health challenges to trauma to deeply ingrained patterns of behavior to systemic challenges that disadvantage the homeless—are localized problems to be solved by the homeless themselves, the residential staff, and others as they attempt to implement—not deliberate over, reformulate, or design—the policy objective presented to them.

Up to this point, Engeström's CHAT has been applied to organizational and cross-organizational collaboration. But public policy is work on a different level, transcending organizations and transforming economies. Tackling problems on such a scale requires considerable translation—translation that to date has been incomplete, as we will see in the rest of this chapter.

HISTORY

To understand how Engeström's CHAT was translated into a method for implementing public policy, let's first review criticisms that had been leveled against Engeström's CHAT throughout the first decade of the twenty-first century.

As mentioned at the end of chapter 8, throughout the 2000s, some criticisms of Engeström's CHAT focused on what was perceived to be a tight, instrumentalist focus on improving inter- and intraorganizational functions. Although Engeström had declared that any capitalist society faced the primary contradiction between exchange and use value (1987/2014, p. 102), critics argued that this primary contradiction was not actually addressed in his work and that capitalism had been insufficiently criticized.

For instance, Elhammoumi (2001) indicts "contemporary theorists" of CHAT, including Engeström, for locking onto small-scale contexts—classrooms, families, work groups—and thus emphasizing "intention, shared meaning, individual or distributed cognition, memory, the development of speech and so forth. This construction overlooks the equally, if not more important, larger scale such as forms of social control and power, distribution of wealth, divisions of labor and social class" (p. 202). He adds that "in their analysis of human cognition, socio-historicocultural psychologists (Bruner, Cole, Engeström, Rogoff, Valsiner, van der Veer, Wertsch among others) kept out of their theoretical picture any reference to larger forms of human activity and larger processes in social life, such as the realities of economic structure, class struggle, realities of labour activities, and the realities of social interaction" (pp. 204–205; note that Engeström is not a psychologist).

Similarly, Avis (2009) argues that although Engeström acknowledges larger social structures,

> this insight fails to be translated into broader political engagement as the focus of his interventions are upon practices within a particular activity system or those adjacent to it, as is the case with third generation AT. Consequently the wider social structure within which these processes arise tends to be ignored. This leads to analysis that leaves these wider relations in place and fails to interrogate the manner in which they shape the terrain on which an activity system or cluster is set. It is this limitation that in part results in analyses and change strategies that effectively *secure the interests of capital*. (p. 156, my emphasis)

So, Avis complains, Engeström's developmental work research focuses on peripheral contradictions rather than primary ones (p. 157). By focusing on peripheral contradictions, "Engeström's AT veers towards becoming a form of comfort radicalism, its transformative rhetoric has a progressive appeal but ultimately it readily lends itself to becoming no more than a management technique" (p. 161). Avis concludes that "there is a tension in Engeström's AT that readily lends itself to appropriation by capital, and is a consequence of its failure to seriously acknowledge social antagonism and wider patterns of social relations in which activity systems are located" (p. 162).

In the same journal issue, Peim argues that what he terms Engeströmian activity theory (EAT) "is restricted by a commitment to a progressive, apolitical ideology of improvement" and "defined in terms of its expression of a will-to-power." It is disengaged from its supposed philosophical tradition, he charges, and thus it has misappropriated Vygotsky's legacy (Peim, 2009, p. 167). He charges that

> EAT, then, has become a widely accepted technology of transformation harnessed to an ethic of improvement. This ethic has beset education as a whole with an overriding logic of performativity. One aspect of this marriage is the insistence, strongly emphasized by Engeström, on the local nature of activity systems. The avowal of a "radical localism" in EAT means that questions about social systems—and about the relations between local practices and larger social systems—are not, and cannot be, addressed (Engeström, 2003, 36). As a result, what is sometimes known as sociocultural historical activity theory gives scant attention to any socially differentiated theory of culture or of history! Only thus can it offer itself as a universal technology of improvement. (p. 168)

Put another way, Engeström's CHAT has no theory of a public. To be fair, CHAT did not need one originally; it started out as a Marxist psychology, studying dyads (chapter 4). Even as it expanded under Leontiev, it was anchored to labor, which was regarded as a sufficient microcosm of society, and deployed apolitically (chapter 5). But as CHAT expanded to sociological concerns (chapter 8), it neglected to develop an account of a public with legitimate, distinct competing interests beyond contradictions.

Still critical, but in a more conciliatory vein, Sawchuk and Stetsenko (2008) note Westernized activity theory's "noncanonical" turn to sociology and argue that it is appropriate, acknowledging that "canonical" activity theory had been prevented from moving into sociology due to the USSR's ban on sociology during the Stalin years (cf. Weinberg, 1974). That is, they note and accept CHAT's translation to an organizational sociology as discussed in chapter 8—at least as a strand (see chapter 3). Still, they urge a more activist stance for CHAT, one that could go beyond organizational functions to address social problems as Vygotsky, Leontiev, and Luria (in their reading) had attempted to do. (Stetsenko amplifies this argument in her 2017 book, in which she urges a transformative activist stance that she characterizes as more similar to Vygotsky's.)

At about the same time that these criticisms were being leveled, Engeström was also thinking about how to address broader social problems. In his 2008 book, he identified *societal problems* such as global warming as "runaway objects." In 2009, he mentioned *social movements* as an example of a phenomenon that CHAT needed to be developed to address (2009b). Also in 2009, in response to his Festschrift (Sannino et al., 2009), Engeström suggested that a fourth generation of activity theory was necessary to better understand social problems: "Third-generation activity theory still treats activity systems as reasonably well-bounded, although interlocking and networked, structured units. What goes on between activity systems is processes, such as the flow of rules from management to workers." But "in social production and peer production, the boundaries and structures of activity systems seem to fade away. Processes become simultaneous, multidirectional, and often reciprocal. The density and crisscrossing of processes makes the distinction between processes and structure somewhat obsolete. The movements of information create textures that are constantly changing but not arbitrary or momentary" (2009a, p. 309).

Like actual mycorrhizae, "social production requires and generates bounded hubs of concentrated coordination efforts," and "activity system models are very appropriate for the analysis of such hubs" (p. 310). Thus, Engeström speculated that a fourth generation of activity theory was needed to better model cognitive trails and mycorrhizae (p. 310).

An early indicator of this iteration's postmarket orientation was Sannino's (2011) article "Activity Theory as an Activist and Interventionist Theory," in which she (like Stetsenko, 2017) characterized the work of the Vygotsky Circle as activist and interventionist. Sannino's examples included

- Vygotsky's work at the Institute of Defectology;
- Luria's early work with the dynamoscope (1960) and his twin studies (Luria & Udovich, 1956/1959);[2]
- Leontiev's work at the hand rehabilitation hospital (Leont'ev & Zaporozhets, 1945/1960).

But to address societal problems, the scope of investigation had to broaden again. Thus, Engeström and Sannino began to investigate social movements

(Engeström et al., 2016) and alternatives to capitalism (Engeström & Sannino, 2016; Sannino & Engeström, 2017). Stung by the early 2000s criticism of Engeström's CHAT as ignoring the primary contradiction of capitalism, in their interview with Yamazumi, Engeström and Sannino pushed back, repositioning Engeström's CHAT to study the "acute societal problems of our time" (Sannino, quoted in Yamazumi 2020, p. 3). This repositioning involved a new scope, as we'll see below, but was founded on the concept of the cycle of expansive learning as well as the change laboratory methodology. As with the later change laboratory work, this translation of Engeström's CHAT to a method for implementing public policy involved less focus on designed mediators, as in chapter 6, and more focus on accepting coordinating genres (such as agreements) that solidify relations. These coordinating genres reflected the broadening scope of analysis and intervention: as the scope of intervention moved from bounded scenes (chapter 6) to interorganizational collaboration (chapter 8) to societal issues (this chapter), the nature of the intervention moved from codesigned individual artifacts and practices (chapter 6) to novel genres codeveloped by participants (chapter 8) to more standardized genres through which the participants' agency was transformed (this chapter).

EMPIRICAL FOCUS: PUBLIC POLICY PROBLEMS

Thus, the empirical focus broadened again, temporally and spatially expanded beyond the interorganizational collaborations described in chapter 8 to address more durable, expansive societal problems. As Sannino says in Yamazumi's interview,

> we are dealing with different types of objects that are qualitatively different from what we have been dealing with so far; therefore, our unit of analysis must also be qualitatively different. One idea I have been playing with is—this is also in line with what Yrjö said—to look at coalitions playing out at different levels. Vertically and horizontally, in time and in space, I have been thinking about when these intersections could be considered as nodes of learning that are like clusters or germ cell formations, which condense the learning process during one phase and open new perspectives for the next phase. (Sannino quoted in Yamazumi, 2020, p. 4)

Such objects are the focus of "heterogeneous coalitions" (Engeström & Sannino, 2021, p. 19) that involve "a wide variety of actors at multiple levels—local, regional, national and possibly global" (p. 14). In one article, their extended example includes actors such as NGOs, housing units, municipal services, state-level actors, and neighborhood associations—that is, institutional actors—as well as staff and clients at a housing unit (Engeström & Sannino, 2021). All of these actors are involved in intertwined learning cycles centered on a shared public policy object (see also Sannino, 2023), one that by its scope is "qualitatively different" from those examined previously (Sannino quoted in Yamazumi, 2020, p. 4). In this translation, Engeström's CHAT broadened its empirical focus further, meaning that a phenomenon either was newly created or became newly visible: the *policy objective*.

To examine the heterogeneous coalitions that cluster around these qualitatively different objects, Sannino and Engeström applied change laboratory methodology, which had previously been applied in more localized cases (such as hospitals, libraries, and banks) to help people deliberate on futures. The change laboratory methodology encourages deliberation, opening up a full range of questions for participants to deliberate on. Mediators allow the participants to intervene at various conceptual levels, from lower-level epistemic questions such as What?, Who?, When?, and How? to higher-level questions such as Why? and Where to? (Engeström, 2007; Engeström et al., 2006; cf. Bødker & Iverson, 2002; cf. the "why," "what," and "how" layers in Bødker, 1997). The change laboratory method naturally generates such questions by encouraging six kinds of participant agency:

- *Criticizing* the current activity and organization.
- *Questioning or resisting* the interventionist or the management. Resistance toward the intervention process or the management may take the form of criticism, questioning, opposition, and rejection.
- *Explicating* new possibilities or potential in the activity. This refers to reflecting on previous positive experiences and seeing them as a potential left unacknowledged or characterizing the problematic object as a source of new possibilities.
- *Envisioning* new patterns or models of the activity. Envisioning can occur in the form of preliminary suggestions or presentation of comprehensive models for the future.
- *Committing* to concrete actions aimed at changing the activity and expressing this through commissive speech acts.
- *Taking* consequential *actions* to change the activity. Participants of the interventions may actually change the situation through a sequence of

actions. This may occur in, between, and after the laboratory sessions (Engeström et al., 2012, p. 404).

In Engeström's CHAT-as-organizational-sociology (chapter 8), when people in change laboratories made their way to these higher-level questions, they questioned the concepts given "from above" by their institution, instead substituting and then synthesizing concepts that emerged "from below" during the change lab (Engeström et al., 2006). Thus, the change lab became a mechanism for these individuals to critique and reformulate concepts rather than taking these concepts as a given. They were empowered not just to *implement* directives "from above" but to *remake* these directives. As Haapasaari et al. (2016) put it, "Transformative agency goes beyond the individual as it seeks possibilities for collective change efforts" (p. 233), and it can involve questioning a problematic object that has been given by management (p. 236). That is, change labs offered workers not just help in working more effectively (the "how"), but a means to *participate in reformulating the object of their work* (the "why" and "where to"—but see Langemeyer & Roth, 2006, for a critique). Thus, change labs seem like a good fit for scaling to public policy participation. But, as we'll see below, the fit is not perfect.

THE INTERMESHING OF MULTIPLE CYCLES OF EXPANSIVE LEARNING

These "heterogeneous coalitions" (Engeström & Sannino, 2021) are understood as entangled in learning cycles, interventionally addressed through change laboratories and cycles of expansive learning (CELs). In theoretical terms, this new iteration focuses on the interplay across entities at different levels and thus multiple cycles. Thus, the unit of analysis moves to the *intermeshing of multiple cycles of expansive learning*, or as Sannino puts it above, "coalitions playing out at different levels." The result is a new type of activity formation. Here is how Engeström and Sannino characterize the problem space to which they apply their new unit of analysis:

> To us, the main challenge that requires a fourth generation activity theory is the radical and fateful transformation in the *objects* of human activity. In the current phase of capitalist globalization, such interconnected objects as poverty, climate change and pandemics cannot anymore be treated as isolated issues to be brought under control by technical means; they influence and pervade the objects on innumerable activities and call for radical revisioning of the ways our societies and lives are organized. (Engeström & Sannino, 2021, p. 14)

And

> fourth generation activity theory should offer a unit of analysis able to grasp a qualitatively new type of activity formation and concerted efforts that can realistically counteract stigma and suffering associated with conditions of deep disadvantage. Such a new step entails the involvement of a wide variety of actors at multiple levels—local, regional, national and possibly global. Such heterogeneous activities are brought together by objects that are directly related to the future of the planet and the humankind living on it, that is, to urgent socio-economic, ecological and humanitarian crises. (p. 14)

Engeström & Sannino portray this type of activity formation as new. Yet from the standpoint of public policy, it is not: public policy has long been addressed by heterogeneous public and private coalitions with differing interests, and it has long faced local compliance challenges of the sort described in the homelessness case. Homelessness is not a new phenomenon (Vygotsky served and studied children who had been left homeless by the Russian Civil War, for instance; Sannino, 2011, p. 580). Nor is this case newly connected to other factors: homelessness has always had many potential contributing factors, which vary across societies and circumstances, and public policy specialists have long analyzed how these factors interrelate. Again, although the problem is not new, it is *newly visible* to Engeström's fourth-generation CHAT: it becomes visible once CHAT widens its unit of analysis beyond interorganizational collaboration.

In chapter 8, we saw that when Engeström's CHAT expanded to cases of organizational sociology, it had to undergo considerable translation. Part of that translation involved incorporating some concepts from other disciplines and creating other concepts in contradistinction to existing ones. These new sociological concepts were necessary for addressing the spatial, temporal, and social expansion in empirical focus that Engeström's CHAT underwent. It arguably has a way to go before successfully integrating these new sociological concepts.

In its turn to public policy in this chapter, Engeström's CHAT is facing even more expansion. One way to handle this continuing expansion would be to follow the same pattern as it did with managerial and social theory: incorporating some concepts and creating others in contradistinction to existing ones in public policy. It's possible that Engeström's CHAT will follow this path in the future—but in the present, this path has surprisingly *not* been followed. In this research, Engeström and Sannino do not appear to be citing public policy research at all.

This creates a significant barrier for the interessement of Engeström's CHAT as public policy. Without engaging in this literature, Sannino and Engeström can't address the relative strengths of this framework compared to the frameworks developed by public policy experts, nor can they develop CHAT concepts in contradistinction to that literature.

And that literature is substantial. It offers theoretical and methodological frameworks for addressing how policies are researched, formulated, validated, piloted, and subjected to review by the public and by affected agencies and entities. Given the breadth of individuals who must be characterized in order to provide evidence for public policy decisions, this research tends to rely on quantitative data, sometimes paired with qualitative data. For instance, in a landmark study of the homeless in my own city of Austin, Snow & Anderson (1993) paired qualitative research (participant-observation) with quantitative research (tracking a sample of homeless through a network of core institutions)—providing qualitative thickness while still allowing them to characterize homelessness at the level of a city. In another study, researchers reviewed 30 years' worth of archival data to investigate the relationship between homelessness and age cohorts in New York City—a relationship that uncovered contributing factors to homelessness over time (Culhane et al., 2013).

Strikingly, Engeström and Sannino (2021) do not describe change labs as a way of deliberating over or improving public policy. Instead, they accept the Housing First principle as a given and then work to find ways to *implement* that principle. In other words, unlike in previous change lab work, the participants in this study were led to address only lower-level epistemic questions such as What?, Who?, When?, and How? They do not appear to have addressed higher-level questions such as Why? and, especially not, Where to? They were not encouraged to question the concept (the Housing First principle) that had been given "from above" and certainly not to rethink, reformulate, or remake the policy. To "enact the shared utopia" (Engeström & Sannino, 2021, p. 14), individuals must accept—not criticize, question, or resist—a common vision of the utopia. They must comply with that single vision, and make choices that help to enact it, bringing them "toward the next steps of the strategy" that has been laid out at the national level (p. 19).

Stepping back from supermediators, Engeström & Sannino have returned to Leontiev's *supervised citizen*.

The central story they offer at the housing unit level is one of *compliance*. Without consulting staff, a new manager decided to demolish the protective wall (a plexiglass window) that had been erected between residents and staffers. Rather than criticizing, questioning, or resisting this decision, staffers were envisioned as accepting it and complying by choosing to sit with a resident rather than staying in safety. It's not clear to what extent staffers actually did this—the language becomes speculative here—but we're told that the decision to remove the wall "led to great turmoil among the staff" because "some of the workers were afraid of the residents, and the idea of being in an open space with them without a protective wall in between was very difficult for them to accept" (p. 17). Indeed, the workers' fears had at least some validity: elsewhere, Sannino quotes a staffer who described aggressive behavior from residents, behavior that sometimes required him to call the police (Sannino, 2018, p. 390).

To convince staffers to take on more personal risk, they draw on the language of *double stimulation*, which in Vygotsky's work was a method for identifying sign mediation (see chapter 4) and in their own previous work had been characterized as a powerful principle for collective intervention underpinning change labs (Sannino & Engeström, 2017). But in contrast to those other implementations, double stimulation is offered here as a *management tool*.

What of the other stakeholders? The residents did not participate in the change labs at all, but we are told that "the residents largely welcomed the change as a perspective toward regaining their self-esteem and self-determination" (Engeström & Sannino, 2021, p. 18). The neighborhood association did not participate either, although we are assured that "the learning taking place within the housing unit directs and support[s] learning in the neighborhoods" (p. 16). Of the other represented stakeholders, we are told that "two related Change Laboratories followed in 2019, one with city level actors in the city of Tampere and the other one with state-level actors" (p. 18), but not who these actors were. The change labs were "intertwined," but in ways that sound minimal: one practitioner from one change lab presented at a second change lab. Two participants from previous change labs participated in a subsequent one. Some video clips from the first two change labs were used in the third. How deeply did the parties interact? To what extent did the carried-over participants represent the work of their previous change labs? How many video clips were used, and how did they impact the deliberations of the third change lab? The answers are not clear; the results are being analyzed.

Other publications on the same project paint a similar picture. The broader scope of public policy involved the more abstract (runaway) object of homelessness, a "wicked problem" (Engeström in Yamazumi 2020, p. 3), but one framed in simplifying dilemmatic language: "living a healthy life in the apartment rented to them vs. continuing to use drugs and alcohol" (Sannino, 2018, p. 389) or "the motive of living 'a life of human being' clashing with the motive of taking amphetamine" (Sannino, 2020a, p. 10). As a housing unit supervisor stated to Sannino, "The important thing is to get the will of the resident expressed," provided that the will is in accordance with the housing unit (Sannino, 2020a, p. 9). In the broader scope, the policy objective is characterized and accepted, and the intervention focuses on persuading people to comply with it.

Change laboratories in this translation of Engeström's CHAT follow two principles: double stimulation and ascending from the abstract to the concrete (Sannino & Engeström, 2018, p. 50).

DOUBLE STIMULATION

As mentioned in chapter 4, double stimulation was originally used by Vygotsky as an analytical tool to examine sign mediation in individuals (cf. Vygotsky, 1931/1997c). But, just as with the zone of proximal development, this technique for analyzing *individual* cognition has been repurposed for intervening in the activity of social *groups*—here, motivating them to accept public policy dictates.

As Sannino explains, double stimulation involves two kinds of stimuli.

- **The first stimulus**: "a key problem participants are confronted with"
- **The second stimulus**: "an artefact or a model, a theoretically inspired tool that would help the participants to redesign their activity" (quoted in Ploettner & Tresseras, 2016, p. 96–97).

In the change laboratory methodology, "double stimulation is not only a method understood as a technique. It's a Method (with the capital letter) understood dialectically as a generative principle that will be instrumental to develop our theory further" (Sannino, quoted in Ploettner & Tresseras, 2016, p. 96–97).

For instance, in the Housing First case:[3]

- **The first stimulus** or key problem for residents is that they face the dilemma of continuing their current destructive behavior or complying with housing rules (Sannino, 2018, p. 389).
- **The second stimuli** include coordination genres such as lease agreements, support contracts, service plans, and letters of intent (Sannino & Engeström, 2018, pp. 74), which build the coalition's agency via local commitments (pp. 74–75). These documents address the dilemma of the first stimulus: "The signed rental agreement and negotiated care plans serve as mediators and supporting agency formation anchors" (Sannino, 2018, p. 389). That is, they are not coproduced (designed mediators, as in chapters 6 and 7), but are ready-made solutions (designing mediators).

But the residents are not the only ones receiving such stimuli: "staff members who had worked for many years acting as guards and controllers" were used to working behind "a plexiglass window through which they could control those who entered or exited the unit." But the Finnish Housing First approach dictated a new role, not as a guard but as a supporter for residents:

1. A new unit manager eliminated this protective plexiglass window.
2. Staffers became afraid of interacting with residents in the open space.
3. Then the unit manager "sat with them in the open space casually interacting with the residents or having oatmeal with them."
4. Through this action, "a set of new mediating means (or second stimuli) became available to the workers" (Sannino, 2020a, p. 8).

Unfortunately, taking down the plexiglass window "terrified" the staff (Sannino 2020b, p. 12), perhaps for good reason, since they sometimes had to call the police to deal with disorderly residents (Sannino, 2018, p. 390; Sannino et al., 2021, p. 10). Shortly before the change lab began, this fear "led to several resignations among the personnel" (Prokopis et al., 2022, p. 5).

The subsequent change lab focused on persuading the remaining staffers to accept this institutional discourse and implement the policy. Using speculative language, Sannino frames the result by characterizing compliance as transformative agency:

> A staff member in this unit *might* have been torn between the conflicting motive of sticking to the old guard-like way of approaching the residents and the new way of working (first stimulus). One day, this staff member *might* have decided that the next time he would meet this resident of whom he is so much afraid, he will ask him if he wants to have a cup of coffee (second stimulus) following the manager's

> example. When the staff member meets this resident, the volitional action *might be implemented or might not*. If the volitional action is accomplished by implementing the second stimulus, the problem situation acquires a new meaning, becoming less paralysing. Then by repeating over and again the process, transformative agency develops to the point that the staff member discovers new capabilities he had ignored before and qualities in the resident that he had equally ignored. (Sannino, 2020a, p. 8, my emphasis)

Sannino *speculates* that the double stimulation of the bowl of oatmeal can help the staffers to internalize the authoritative discourse about how to treat the residents—just as the residents themselves are expected to internalize and repeat back the authoritative words of a housing counselor (Sannino, 2020b, p. 10).

Earlier, we saw that critics had characterized Engeström's CHAT as being a management tool. Engeström and Sannino responded by translating it into a fourth-generation CHAT oriented to developing alternatives to capitalism—but at least in this instance, the result seems even more nakedly managerial.

ASCENDING FROM THE ABSTRACT TO THE CONCRETE

"Ascending from the abstract to the concrete," Sannino and Engeström tell us, "is a method of grasping the essence of an object by tracing and reproducing theoretically the logic of its development, of its historical formation through the emergence and resolution of its inner contradictions" (Sannino & Engeström, 2018, p. 51). It is grounded in "a germ cell abstraction," which "captures the smallest and simplest, genetically primary unit of the whole functionally interconnected system"; from this abstract starting point, "ascending from the abstract to the concrete is achieved through specific epistemic or learning actions. Together these actions form an expansive cycle or spiral" (p. 51).

Sannino and Engeström provide the example of mobility as studied in their previous work (Engeström et al., 2012). "We analyzed the formation of the theoretical concept of sustainable physical mobility among the home care clients and workers," finding that "the concept was not yet there. . . . The concept emerged as the practitioners and their elderly patients worked with practical mobility exercises and obstacles to their accomplishment" (p. 52). Importantly, they identified the germ cell, "a recurring item that seemed to function as a key that opened up pathways to increasingly independent forms of mobility for the elderly": "the simple action of standing up from a chair. . . . We identified it as the germ cell of the emerging concept of sustainable mobility. This germ cell is a contradictory unity of opposites: the quest for safety on the one hand, and the quest for autonomy on the other hand" (p. 52; cf. Engeström & Sannino, 2016).

In the Housing First case, the germ cell is the principle of Housing First—that is, providing housing *before* asking the client to modify her or his behavior (Yamazumi, 2020).

MASTERY: OF POLICY OBJECTIVES

In Engeström's CHAT-as-organizational-sociology, the focus was on collaboratively producing organizational and cross-organizational agreements. That version of Engeström's CHAT asked functional representatives to generate solutions. But in Engeström's CHAT-as-public-policy-approach, as the empirical focus widened to public policy questions, such an outcome becomes hopeless: public policy is not generated by functional representatives working in service institutions, but by political representatives who work in legislatures and government agencies.

In public policy, which is focused on complex, durable societal problems with multilevel, multitype coalitions, who learns to master the activity? That is, where is agency located? As implied by our earlier discussion, agency is distributed across activities. But that does not mean that everyone participates in every decision. In the homelessness case, as we've seen, the individual agency of staffers and clients is transformed as they accept policy objectives and enact ways to accomplish those objectives. In practice, this means that individuals are positioned as enacting or accepting given genres (such as agreements and treatment plans). To "enact utopias," individuals must accept a common vision of the utopia, comply with that vision, and make choices that help to enact it—such as residents choosing to remain in an apartment rather than continuing to abuse alcohol, or staffers choosing to interact with residents rather than staying behind a wall.

This vision of mastery—as accepting supervision—parallels Leontiev's managerial vision as expressed in his work at the rehabilitation hospital, in which soldiers' rehabilitation was framed as being oriented to the state's goals. (Like Sannino, the USSR also urged its citizens to enact a utopia, the utopia of communism.) But it contrasts with the earlier visions of mastery in Westernized CHAT. In those previous interventionist efforts, participants

- "question[ed] the validity of tasks and problems posed by the context and [began] to transform the context itself" (Engeström, 1994, p. 17) through criticisms guided by the cycle of expansive learning and amplified by designing mediators (mirror materials and models);
- criticized current tools and codesigned their own (Bødker, 1991) through designed mediators such as mockups and prototypes;
- collaboratively developed new coordinative artifacts to solve a problem in ways that met individual as well as institutional needs (Nummijoki & Engeström, 2013) through collaborative work on designed mediators such as draft agreements.

That is to say, they *deliberated*, engaging in dialogue as they attempted to identify an emergent solution that would meet their diverse needs. In the CHAT-as-public-policy-approach, participants instead accept criticism, self-criticize, enact official genres, and comply with the given state policy objective, mastered by the rhetorical power of the change lab rather than enforced by the police power of the state.

Readers, don't misunderstand my critique here. I am not saying that national policies should be deliberated or codeveloped by the individual staffers who must implement them. That would make national policies impossible to implement. Yes, staffers must comply with the policy directives they're given in order to make programs work, and it usually helps if they believe in (i.e., do not question, reformulate, or try to improve) the policy objective that they are given. If change labs can help that implementation work, that's an important contribution. But it is a contribution that unwinds CHAT's translations, turning it back to Leontiev's supervisory stance.

MEDIATORS: SECOND STIMULI AND COORDINATION GENRES

Finally, mediators undergo a slight but significant change in this version of CHAT. CHAT-as-public-policy-approach uses the same two classes of mediators noted in chapter 8's discussion of change labs:

- **Second stimuli**: Mediators used in "double stimulation," such as mirror materials and conceptual tools.
- **Coordination genres**: Genres such as agreements and leases. These are similar to the care agreements, mobility agreements, rental agreements,

negotiated care plan documents, and other genres developed for change laboratories in Engeström's CHAT-as-organizational-sociology.

However, in CHAT-as-public-policy-approach, the coordination genres are no longer *designed* mediators; they are *designing* mediators. They are oriented to the policy objectives of the coalition, interacting with the interlocking cycles of expansive learning of that coalition. Instead of producing a designed mediator, they produce transformative coalitional agency. As we saw, it is taken as a positive sign when a resident *internalizes and repeats* the authoritative discourse of the housing counselor (Sannino, 2020b, p. 10) and when the staffer in turn *internalizes and lives out* the authoritative discourse of the manager (2020a, p. 8).

SUMMARY: TRANSLATING CHAT TO IMPLEMENT PUBLIC POLICY

This new translation from organizational sociology to implementation of public policy is nascent. In contrast to the organizational sociology in chapter 8, which engaged deeply with the extant non-CHAT literature, this translation has not engaged with the literature of public policy. Perhaps it is because this translation is so recent, but the analyses so far have been less elaborative, less expansive, and more didactic.

In taking up CHAT in this iteration, Engeström and Sannino translated it in key ways:

- It added still more stakeholders, representing not just activities but heterogeneous coalitions representing the entire public.
- It reflected the role of ground-level public policy efforts, which is to implement political decisions made by political representatives rather than to develop new policies from the ground up.
- Thus, in resisting criticisms about Engeström's managerial discourse, it engaged in still more managerial discourse. Change labs, which had previously been deployed to design new mediators, were now deployed for purposes of transformative agency—which in concrete terms meant persuading participants to accept state supervision and the associated authoritative discourse.

These changes can be summed up in three different translations:

- **Empirical focus**: Its empirical focus widened from interorganizational collaborations to public-scale institutional coalitions implementing public policy objectives such as homelessness, poverty, climate change, and pandemics.

- **Mastery**: Its understanding of mastery shifted from *generating* an organizational agreement to *accepting and implementing* public policy decisions.
- **Mediators**: As the empirical focus widened, its mediational focus shifted from designing coordinative genres to transforming agency.

Through these translations, Engeström and Sannino

- **problematized** public policy initiatives that faced problems in ground-level adoption;
- **interested** public officials and agencies, public workers, the homeless themselves, and communities in which the homeless were housed, making their arguments to institutional funders and CHAT scholars—including the CHAT scholars who had earlier questioned the Marxist bona fides of Engeström's CHAT;
- **enrolled** actors such as the Finnish Housing First initiative by seeking and receiving institutional funding;
- **mobilized** all of these enrolled actors, expanding the change laboratory approach by describing chained change laboratories in which the results of one change lab served as partial inputs to the next. This mobilization resulted in the transformative agency of the coalition—that is, in policy compliance.

Engeström and Sannino translated CHAT-as-organizational-sociology into a method for implementing public policy, one that could address larger "wicked problems" (table 9.1). Will this settlement hold? We are seeing some uptake of this translation in education and other public policy concerns, but it's too early to tell how successful it will be.

Despite the drawbacks discussed above, we may be able to see potential in applying change labs and other interventionist CHAT approaches to public policy issues. Let's end with one more illustrative case for CHAT as an implementation of public policy: the case of organic agriculture in Zimbabwe (Mukute et al., 2018).

In a strand of research that Engeström and Sannino cite to illustrate this translation of CHAT (Yamazumi, 2020; Engeström & Sannino, 2016), Mutizwa Mukute, Heila Lotz-Sisitka, and colleagues at Rhodes University have examined cases such as "(1) sustainable agriculture in Lesotho; (2) seed

Table 9.1
Domain, project, translations, and trials in Engeström's CHAT-as-public-policy-approach

Domain	Public policy
Project	Acute societal problems; alternatives to capitalism
Translations	
• ***Empirical focus***	Public policy objectives, instituted in public-scale institutional coalitions
• ***Mastery***	Inducing individuals to accept and implement public policy decisions
• ***Mediators***	Charter documents to reflect public policy implementation
Trials	To be determined

saving and rainwater harvesting in Zimbabwe; (3) community-based irrigation scheme management in Mozambique; and (4) biodiversity conservation co-management in South Africa" (Lotz-Sisitka et al., 2017, p. 897; cf. Colvin & Mukute, 2018; Mudokwani & Mukute, 2019; Mukute, 2016; Mukute & Lotz-Sisitka, 2012; Mukute et al., 2018; Vogel et al., 2021). In these related studies, the authors examine how "expansive learning might also facilitate instances of transgressing norms—viewed here as embedded practices which need to be reframed and changed in order for sustainability to emerge" (Lotz-Sisitka et al., 2017, p. 898). Broadly, these projects focus on decolonizing agriculture, noting that colonialism had undermined periodic solidarity, leading to the loss of common vision, new practices that undermined solidarity, an absence of relationship-building mechanisms, and differential treatment (Mudokwani & Mukute, 2019). These studies have been supported by funding bodies such as Oxfam (Colvin & Mukute, 2018), UNESCO (Mudokwani & Mukute 2019; Mukute et al., 2018), and the South African Qualifications Authority (Mukute & Lotz-Sisitka, 2012).

Mukute et al. (2018) specifically examine organic agriculture in South Africa. They state that CHAT points to "the limitations of current problem-solving approaches that have been developed in and tend to serve a capitalist-based approach, which commodifies knowledge, natural resources, and life forms" (p. 244). In their interventionist research with "eight interacting district organic farmer associations" across Zimbabwe (p. 233), they conducted four change laboratory sessions. The first session, involving ninety-nine organic farmers and ten extension workers, was conducted over eight days and focused on identifying challenges and generating mirror data for the later sessions. The second through fourth sessions involved thirty-nine farmers and seven content specialists and focused on analyzing mirror data and then modeling and refining solutions (p. 235). The fourth section, in particular, involved three groups presenting solutions to each other and critiquing these solutions (p. 236).

Unlike the Housing First studies, then, these change labs sessions were focused on developing localized policy solutions rather than accepting public policy decisions.

Through these discussions, the interventionist researchers identified how "the matters were not only interconnected but also stratified":

> For example, climate change causes water shortages through droughts and longer midseason dry spells, and water shortages undermine food production, which results in food insecurity and poverty. Research participants' analyses of the matters of concern using problem tree analysis suggested that the nexus issues such as food insecurity, water, and climate change had multidirectional causal relationships. For example, poverty and food insecurity make farmers vulnerable to climate change, and climate change weakens farmers' abilities to produce food and get out of poverty. (p. 237)

A fifth change lab examined why and how the solution was locally successful (Mudokwani & Mukute, 2019, p. 6).

Through this change laboratory approach, Mukute et al. argue, the participants were able to participate in *transgressive learning*, which challenges "unjust and unsustainable norms and practices that have become normalised" (p. 229). In transgressive learning,

> transformations to sustainability may involve transgression of unjust and unsustainable norms. Such transgression and associated transgressive learning are interested in protecting and caring for the earth and life in and on it. . . . Such learning can be supported through the CHAT-informed expansive learning process using the CL method. However, the current CHAT conceptualisation of goods and services that are produced, exchanged, and distributed seems to exclude common good and ecological services that are important in transformations to sustainability. (p. 245)

These studies set much more modest goals than the Housing First project, characterizing their work as generating insights from the concerns of local farmers' associations (Mukute et al. 2018, pp. 243–244). Yet, rather than focusing on compliance with public policy, these studies focused on how transgressive learning from the local level could impact "unjust and unsustainable norms" more generally (p. 245). That is, they set out not to *implement policy objectives* but, rather, to explore local insights that could *feed into policy planning*.

Still, it's worth noting that Mukute et al.'s study was itself funded by, and localized policy solutions presumably had to align with the policy preferences of, the NGO and UN sponsors of those projects. Even with its focus on transgressive learning, this study was not purely local.

HOW WELL DID ENGESTRÖM'S FOURTH-GENERATION CHAT WEATHER ITS TRIALS?

In the previous chapters, I was able to end by describing a settlement—even if it was an unstable one—that the translation had yielded. But in this chapter, I can't: CHAT-as-public-policy-approach exists as a sketch or a proposal, but not yet as a settlement. It has not yet been subjected to the trials of sustained contact with rivals (the public policy literature), and thus it remains to be seen whether such a settlement can be developed and whether it will hold.

As it currently stands, in translating CHAT to a method for implementing public policy, Sannino and Engeström widened its scope again. At this much broader scope, agency is redistributed. In the Housing First case, *the "should" becomes stronger as the object becomes larger*. The broader the empirical scope, the harder it is to emphasize individual agency and critique over more abstract collective objectives. Individual agency seems to evaporate or disperse across the coalition, and the methods have not been changed to address the new scope. The deliberation-and-design aspect that was so strongly emphasized in Westernized CHAT and CHAT-as-organizational-sociology (chapters 6 and 7) has been reduced to accepting authoritative discourse and finding ways to implement authoritative policy objectives. Yet, as Mukute and his collaborators demonstrate, CHAT can also be deployed in ways that keep its radical localism, yielding qualitative data that can feed into public policy approaches.

Engeströmian CHAT must attend to how theory and methodology scale together. How might it do that? In part III, we will take a step back to contemplate the history of translations we explored in part II. Specifically, we will examine how these translations have yielded broad changes in CHAT's scope, in terms of theory, methodology, and units of analysis.

III TRANSLATIONS OF SCOPE

10 THE SCOPE OF THE EXPANDING OBJECT

With the background in part II, we can revisit the gulf between the two triangles we discussed in chapter 1: Vygotsky's depiction of mediated action (figure 1.1) and Engeström's depiction of an activity system (figure 1.2). This gulf is not just one of different authors but—as we saw in part II—different scopes of empirical investigation, as CHAT was translated and redeployed to address individuals in society, then dyads, then small groups, then organizations and interorganizational collaborations, and finally interorganizational workings in the public sphere. These scopes are critical to understand as we consider how to apply CHAT to workplace studies: As momentary scenes such as doctor-patient consultations (Engeström, 1990, 1992)? As a longitudinal study of medical care (Engeström, 2000)? As distributed, intermittent work (Engeström 2009b)? As multiinstitutional cycles of expansive learning (Engeström & Sannino, 2021)?

In being repeatedly translated, CHAT has expanded its scope to address different scenes of work. In expanding its scope, CHAT has had to address more perspectives, cycles, and recruited activities, complicating the theoretical and methodological apparatus. Put another way, CHAT has been deployed as a theory of context, and there is always more context—leading CHAT to necessarily expand the scope of its investigations (see table 10.1).

This trajectory has been indexed to the *object*, around which an activity forms. As objects expand, becoming broader in scope and more representational, they require broader analytical scope to address them. In this chapter, I examine this methodological and theoretical expansion of the object,

Table 10.1
Domain, project, translations, and trials in CHAT

	Chapter 4	Chapter 5	Chapter 6	Chapter 8	Chapter 9
Domain	Psychology, pedology, defectology	Psychology	Design	Organizational sociology	Public policy
Project	The New Human	The supervised Soviet citizen	Emancipation of labor	Reorganization of interorganizational work	Acute societal problems; alternatives to capitalism
Translations					
• ***Empirical focus***	Sign mediation in individual and dyadic interactions	Tool-mediated labor activity	Deliberations about labor activity in larger groups	Organizations and inter organizational collaborations	Public policy objectives, instituted in public-scale institutional coalitions
• ***Mastery***	Exercising self-control	Accepting state supervision	Participating in the design of mediators	Generating an organizational agreement	Inducing individuals to accept and implement public policy decisions
• ***Mediators***	Physical and psychological tools	Crystallized labor and division of labor	Rules; codesign of mediators	Charter documents to reflect organizational agreements	Charter documents to reflect public policy implementation
Trials	Stalinist orthodoxy; the increasingly supervised Soviet citizen	Stalinist orthodoxy; tension between the theoretical and empirical; the status of activity	Abbreviated scope; changes in work; stable sets of interlocutors; the status of activity	Stable sets of interlocutors; partial integration with managerial and social theory; boundaries of activity; instrumentality	To be determined

revisiting the series of translations that CHAT underwent on the way to the current maximalist scope of Engeström's CHAT.

At the end of this chapter, I'll provide my own triangle to illustrate this expansion.

THE OBJECT AS THE SENSE-MAKER AND THE AMBIGUITY OF THE OBJECT

For CHAT, the object is the linchpin of the analysis, particularly in its relation to the outcome. Kaptelinin called it "the sense-maker," which "gives meaning to and determines values of various entities and phenomena" (Kaptelinin, 2005, p. 5). Engeström et al. (2003) declared that "the object determines the horizon of possible actions" (p. 152). In the terms used in case study methodology, *the object bounds the case* (Yin 2003, p. 22; Creswell 2006, pp. 97–98): in identifying an object, the researcher identifies the other parts of the activity system that are arrayed to transform it (see chapter 11). In more colloquial terms, the object is the "text" to which the rest of the activity system is the "context," the principle by which the analysis is delimited.

In this conception, the object is both objective and projective: both the raw materials and problems at hand and the use value that is envisioned for those raw materials and problems (Engeström, 1990, p. 79). For example, the object of construction includes both the construction materials and the building they will form (Miettinen & Virkkunen, 2005, p. 444; cf. Kaptelinin, 2005, p. 5; Nardi, 2005). The brute earth becomes a field of grain (Engeström & Escalante, 1996, p. 360); the raw metal becomes an implement (Engeström, 1992, p. 107); the fish becomes dinner (Leont'ev, 1975/1978, p. 63). The object defines what counts as an activity for a particular analysis (Miettinen & Virkkunen, 2005, p. 444), and in doing so, it delineates "real activities realised by identifiable people in identifiable locations" (Engeström, 1987/2014, p. 250).

And yet, despite its function as the linchpin for the CHAT analysis, the notion of the object is rather ambiguous when applied to phenomena that are less material and less concrete.

Even in 1987, Engeström mixed his closely bounded examples and cases with examples of extraordinarily broad, relatively unbounded activities such as "science," "learning," and "vagabondry" (1987/2014). As Engeström (1990) stated, "contexts are activity systems" (1990, p. 79)—and context is often treated as the opposite of a bounded case: an expansive horizon.. There's always more context.

Yet when Engeström applied CHAT to his early concrete case studies, the results were tightly bounded. For instance, his 1990 studies included specific classroom and workplace activities with defined, concrete objects that delineated activity systems: a classroom unit about eclipses, physicians' consultations, record keeping at a hospital, specific instances of children's play. Similarly, his 1992 studies described activities such as courtroom judging and medical consultations performed by multiprofessional teams. In each of these cases, Engeström could articulate a particular object—and sometimes conflicting objects—around which an identifiable cyclical activity could be discerned. These activities include *primary medical care*, whose object is the patient (Engeström, 1987/2014; Engeström, 1990, p. 109); *health care*, whose object is the consultation (1990, pp. 71, 85); and *judging* in driving-while-intoxicated (DWI) cases, whose object is the defendant and the case file (1992, pp. 29, 61).

After 1992, though, the object began to expand and diffuse, even in case studies, as Engeström's CHAT transitioned into an organizational sociology (chapter 8) and an implementation of public policy (chapter 9).

THE EXPANSION OF THE OBJECT: DIFFUSING FROM THE CONCRETE TO THE ABSTRACT

As Engeström's CHAT has been applied to broader scope and to more complex and interdisciplinary activities, the object itself has become harder to describe in cyclical terms. In a complex collective activity, identifying a single object is difficult. An activity's participants may have different motives (Hyysalo, 2005, p. 22; Nardi, 2005, p. 40) and different perspectives on the shared object (Christiansen, 1996; Engeström, 1999b; Foot, 2002; Holland & Reeves, 1996). In any collective activity, an object is "multifaceted, evolving" and even "dialogical" (Foot, 2002, pp. 132, 138), understood differently by different participants at different points, developing over time.

The object becomes even more difficult to identify in complex work in which the activity crosses field, trade, and disciplinary borders (Engeström, 1995; Engeström et al., 1995; Tuomi-Gröhn & Engeström, 2003). Activities become polycontextual: "knowledge of procedures is incomplete, problematic situations change constantly, and solutions are typically of short duration," so "continuous experimentation in ongoing production becomes the new basis of expertise" (Engeström et al., 1995, p. 331). Activities often involve co-configuration, in which objects expand both spatially and temporally

(Engeström et al., 2003, p. 154). Activities become polymotivated, with stakeholders acting on the object to satisfy multiple, sometimes conflicting motives (Kaptelinin, 2005, p. 13; Nardi, 2005). Stakeholders become more heterogeneous (Miettinen, 1998), as does the object itself (Miettinen, 2005).

These complications, which result from the repeated translations in part II, perhaps make our understanding of the object richer. But they also challenge the notion of making the object the linchpin of a CHAT analysis because the object can only be identified by exploring perspectives (Foot, 2002), something that can happen only after the unit of analysis has been selected and the investigation has begun. In fact, these complications make it difficult to understand activity as unitary at all and, in turn, have led to conceptualizing activity networks in which multiple activities overlap the same spaces, actors, and tools, often presenting contradictory objectives (Miettinen, 1998, 1999; Saarelma, 1993). That is, by expanding the object, we lose the boundedness that made early CHAT case studies so defined—and that made the activity system a viable unit of analysis.

Why is the object expanding? This expansion is driven by two separate but interrelated problems related to the translations of Engeström's CHAT. The first problem is *methodological*: Engeström's CHAT is being applied more and more broadly, to more broadly scoped cases. The second is *theoretical*: Engeström's CHAT is becoming more multidimensional, as more and more human activity is not fitting the past characteristics to which CHAT was developed as a response, and as Engeström's CHAT has steadily expanded its scope.

THE METHODOLOGICAL PROBLEM: MISSION CREEP

Earlier, I noted the different ways in which Engeström's CHAT treats the object in methodological terms: as a sense-maker, as a component of a predefined unit, as a component with unclear relations, and as undefined. Here, I examine key texts in Engeström's CHAT to see how these different treatments emerged. Specifically, I examine how each translation yields a remarkable spatial, temporal, and social expansion of the object, starting with a child's momentary use of voluntary attention and ballooning to a years-long public policy initiative covering an entire nation.

As we saw in chapter 3, Bakhurst (2009) argued that in the West, Engeström's CHAT is principally used as "an empirical method for modeling activity systems" (p. 197). Problematically, Engeström's CHAT is being applied to more broadly scoped cases, and the object expands accordingly. This problem is prefigured in Engeström's (1987/2014, 1990, 1992) early work. For instance, Engeström (1990) argued that "the entire activity system [is] the unit of analysis" (p. 77) and that "contexts are activity systems" (p. 78). How much is covered by "contexts" is unclear, though, since Engeström stated that an activity system is impinged by other activity systems and "connected to other activity systems through all of its components" (p. 84; cf. Engeström, 1987/2014, chapter 2; 1992, p. 13). That is, like any system, where to set the boundary is an analytical choice rather than inherent in the phenomenon; the boundary is in the map, not the territory.

These connections across activity systems are not just conduits. Engeström (1992) demonstrated that the object itself is beset by contradictions. In the activity of judging, the object is the case file—*and* the defendant it represents (p. 60). In medical consultations, the object is the patient herself—*and* the pain the patient is feeling (p. 109). As Engeström said, "an activity system is by definition a multi-voiced formation" (1992, p. 17; cf. Engeström, 1996b).

Consequently, since the object is not narrowly defined in methodological terms in this early work, it can be expanded. We can already see, for instance, how the objects described in those examples—case files, defendants, patients, pain—do not exist in just one context. Perhaps that is why Engeström (1987/2014) suggested that we seek phenomenological insight before delineating the activity system.

Yet Engeström's later work edges away from bounded activities toward public-sphere issues such as global warming, narcotics, and homelessness—a difficult move, since Engeström's CHAT has known limitations in public sphere issues (see Spinuzzi, 2008, pp. 42–43, 72–73). Public sphere issues are by definition not well contained: they involve entire publics, balance complex competing interests, and do not necessarily lend themselves to a theory of learning and development.

We can see the expansion of the object, and the activities that it defines, through progressive methodological concepts developed in the literature of Engeström's CHAT.

METHOD-MOVEMENT 1: AN ACTIVITY SYSTEM IS FORMED AROUND THE OBJECT

As we saw in chapter 3, the notion of the object defining the activity system, and of the activity system as the essential unit of analysis, appears as early as

Engeström 1987/2014 (chapter 5), where he urges us to delineate an activity system based on an object that is deduced through initial phenomenological work. The model for a single activity system is represented in figure 1.2. Here, the object allows us to identify actors who transform the object, instruments they use to transform it, communities involved in the transformation, and rules and divisions of labor that mediate between the community and the actors and object, respectively. The boundary of the case, therefore, is delineated in terms of the object and its transformation. Methodologically, this boundary is scalable, depending on the scale of the object. That is, an activity system could bound the work of a pair, a small group, a unit, or an organization.

METHOD-MOVEMENT 2: TWO ACTIVITY SYSTEMS SHARE AN OBJECT

Yet, Engeström (1996b) claimed, the minimal model for Engeström's CHAT is not one but two interacting activity systems (p. 133; see also 1987/2014, 1990; see figure 1.3f). In a move that distinguishes his CHAT from Leontiev's (Leont'ev, 1975/1978, 1959/1981), Engeström (1987/2014) introduced this minimal model through the notion of a shared object. He applied Ilyenkov's (1977, 1982) theoretical work with contradictions, specifically contradictions that arise when two activity systems share the same object, to Leontiev's activity theory (Engeström, 1987/2014, 1990, p. 128). For instance, Engeström (1990; 1996b) provided the example of doctors and patients, each with their own activities, sharing the object of the patient. This model of two activity systems sharing an object dominates many of Engeström's early case studies, which feature dyadic encounters such as patient consultations. And with such a contradictory object, the activity must be seen as dialogic or multivoiced. Engeström (1987/2014, chapter 4; 1990, p. 54; 1992, p. 7) began referring to dialogism and multivoicedness early in his work, particularly in terms of expertise.

By introducing the interaction between activity systems—particularly the critical notion of contradictions between them—Engeström (1987/2014, 1990, 1992) immediately widened the context we must examine in order to understand activity. Perhaps "contexts are activity systems" (1990, p. 78), but without understanding the contradiction that is introduced when two activity systems meet, we cannot really understand activity per se. The methodological scope widens.

METHOD-MOVEMENT 3: NETWORKS OF ACTIVITY SYSTEMS SHARE COMPONENTS

In Engeström's CHAT, activity systems can be related in ways other than through shared objects, as seen in chapter 6. For instance, when one activity system takes a particular product as its object (e.g., a company develops software), that object can then become a tool in some other activity (e.g., other organizations then use that software to process their own orders). We might see such activity systems as atomic components that interlink to form larger, but relatively coherent, activity networks (see Miettinen, 1998, 1999; Miettinen & Hasu, 2002). Like the concepts in the first two method-movements, the concept of networks of activity appeared early on in Engeström's work (1987/2014; 1990, p. 13). In fact, we can understand the concept introduced in method-movement 2 (two activity systems sharing an object) as a simple version of an activity network (see figure 1.3g).

In an activity network, interlocking activities can "potentially destabilize each other through their exchanges and inter-penetrations" (Engeström, 1990, p. 13). We have already seen that contradictions in a shared object might destabilize activities. But such contradictions can occur at any point: for instance, an activity system might use an instrument (e.g., software) that conflicts with other tools or other parts of the activity (Spinuzzi, 2003b). So Engeström's (1987/2014, 1990, 1992) early emphasis on multivoicedness becomes quite important here. If the elements of the activity system are truly multivoiced and multivalent, activities do not simply link or collide; they collapse into interpenetrated, overlapping activities that cannot be easily separated (Engeström, 1995; Saarelma, 1993; Spinuzzi, 2003b, 2008).

Consequently, in this method-movement we begin to see work that emphasizes multiple perspectives (Christiansen, 1996; Holland & Reeves, 1996), multivoicedness and dialogue (Engeström, 1995), and polycontextuality and boundary crossing (Engeström, 2001, 2004; Engeström et al., 1995; Tuomi-Gröhn & Engeström, 2003)—all work that reacts to the expansion toward multiple activity systems and the destabilizations that result. Once again, without understanding these multiple potential destabilizations across activity systems that are linked—and later interpenetrated—in an activity network, we still cannot really understand activity per se. Again, the methodological scope necessarily widens, this time with no limit: we could continue traversing the links in the activity network forever.

In such activity networks, activity systems are still delineated as cyclical activities with material objects, so we might expect to be able to arbitrarily select a stopping point in order to bound a case and conduct an analysis. Unfortunately, even this limiting factor begins to erode because we cannot necessarily determine what constitutes a multiperspectival object. In a courtroom, for instance, is the object the defendant or the case (Engeström, 1992, p. 60)? Some objects seem bigger or less bounded than others, and when multiple disciplines are involved, participants may not conceive of the same object in the same terms (cf. Mol, 2002). To conceptualize shared work as being performed on the same object, we must define the object expansively enough to be recognized across boundaries. Yet in this method-movement, we still have the anchor of shared cyclical activity to ground our analysis.

METHOD-MOVEMENT 4: RUNAWAY OBJECTS ARE TRANSFORMED VIA KNOTWORKING IN SUBSTRATES OF MYCORRHIZAE

In method-movement 4, we lose that anchor as well. To understand how, we must dissect the longish title for this method-movement. As we saw in chapter 8, mycorrhizae represent relatively durable horizontal connections across activity systems (Engeström, 2008, pp. 228–229, 2009a, p. 309). These connections allow knotworking, or irregular, distributed pulsing collaborations that may involve different people in different locales during each pulse (Engeström et al., 1999; Engeström, 2008).

Knotworking in mycorrhizae is oriented toward a runaway object, one that is shared by multiple activities with variable actors occupying different locations and collaborating irregularly. And that is how we get from a concrete, material object to an object that is so expansive that it represents a global issue such as global warming, one that potentially involves all human beings on the planet as subjects. The object is no longer common, distinct, and atomic but multiperspectival and runaway.

METHOD-MOVEMENT 5: RUNAWAY OBJECTS ARE TRANSFORMED ACROSS CYCLES OF EXPANSIVE LEARNING

As we saw in chapter 9, to address runaway objects, the bounds of the case had to expand further—to interlinked cycles of expansive learning occurring "at multiple levels—local, regional, national, and possibly global" (Sannino, 2020a, p. 5). The triangular representation of activities drops out, as do the objects themselves. Rather, heterogeneous activities, which function at different levels of society, collectively develop and transform a shared runaway object such as homelessness: an ongoing, complex problem that can only be solved by sustained learning engagement across those heterogeneous activities over years or decades. For this sort of problem, intervention can happen through sustained learning and coordination among participants representing a vast number of stakeholders within these component activities. Sannino and Engeström describe a series of longitudinal change laboratory interventions: "one in a housing unit, one with city-level actors, and one with state-level actors" (Sannino, 2020a, p. 9; cf. Sannino, 2018; Sannino & Engeström, 2018; Engeström & Sannino, 2021).

Up to this point, these change laboratories have been conducted separately, but additional change laboratories are planned at the city and national levels, change laboratories that will "promote mutual learning across sectors, organisations, and levels" (Sannino, 2020a, p. 13) as these different stakeholders attempt to address the mutual object of homelessness. As Sannino stated elsewhere, "What does a homeless person need? A home. What do you need to actually make a difference in the homelessness field? You need a new concept of homelessness, which is a concept centered around the object of this activity" (quoted in Yamazumi, 2020, p. 12).

As mentioned in chapter 9, as a coalition becomes larger, holding it together becomes more difficult. Hence, Sannino appeals to the settled Housing First policy, which warrants narrowing the brief of change laboratories to finding ways to implement the policy, not to question or criticize it.

SUMMARY OF THE METHODOLOGICAL PROBLEM: FROM WELL-BOUNDED STUDIES TO INTERACTING CYCLES OF EXPANSIVE LEARNING

This string of methodological translations represents a feedback loop. The more that Engeström's CHAT researchers recognized how an activity overflows its contextual bounds, the more they expanded the object to help encompass the activity to contain the overflow. Engeström's CHAT aims to be a theory of context, and context is infinitely expandable. Here, we run into the classic methodological problem of how to bound the case (e.g., Yin, 2003). The Engeström's CHAT framework becomes a totalizer, attributing even spontaneous chance interactions to subterranean mycorrhizal roots. To put it colloquially, with this hammer we have a mighty analytical tool, but we also begin to see a world composed entirely of nails.

THE THEORETICAL PROBLEM: CHANGES IN WORK ACTIVITY

The methodological problem hints at the theoretical one. Through the many translations leading to Engeström's CHAT, the object becomes more spatially, temporally, and socially expansive—and thus more shared, contested, and multiple. Context expands, activities interpenetrate, and their connections become more multiperspectival for a reason: work activity itself is changing (Castells, 2003; Guile, 2010; Spinuzzi, 2015a). Changes in work activity have made the object more *multidimensional*: more broadly circulated, shared, and interpreted in different activities. Such objects tend to be representations of dialogic knowledge.

The foundational ideas of activity theory came of age during the industrial era, grounded in Marx's (1990) critique of early industrialization, and developed during the rapid industrialization of the Soviet Union (see especially Luria, 1976). Early activity theorists used examples grounded in agricultural and craft labor: hunting, fishing, farming, blacksmithing. But Engeström (1990, 1992) recognized that work organization is changing in "the age of information technology" (1990, p. 50) and that "there is a pervasive historical transition taking place in the very core of expert cognitive procedures" (1992, p. 10). He argued that we are undergoing a historical transformation in the nature of expertise, moving toward "multi-professional team and network work and expertise" (1992, p. 25). Here, Engeström embraced the hypothesis that society is moving from hierarchy and market to network forms of organization (p. 26; cf. Engeström, 2005), quoting from W. W. Powell. That is, Engeström applied CHAT to knowledge work. And knowledge work, which is primarily mediated by layers of information and communication technologies, is quite different from the primarily tool-mediated agricultural work and craft work that supplied the early illustrations of Engeström's CHAT.

Knowledge work involves creating, manipulating, analyzing, and transforming symbols (cf. Spinuzzi, 2007). Its objects tend to be texts and other representations. Such work has accounted for an increasingly large share of the developed world's economy in the last century (Beniger, 1989; Drucker, 1994; Spinuzzi, 2015a). Even in his early studies (1987/2014, 1990, 1992), Engeström was studying professional work whose objects were primarily representational: in health care and judging, for instance, objects included both people (patients, defendants) and the texts that represented them (computerized records, case files). It's not a coincidence that Engeström's CHAT has become widely adopted by scholars in fields that involve knowledge work: human-computer interaction, computer-supported cooperative work, education, medicine, organizational studies, and writing studies.

Knowledge work's objects are transformed differently from objects of agricultural and craft work, typically drawing on multiple textual resources. Agricultural work and craft work tend to have concrete objects and often relatively simple divisions of labor. Hunting, fishing, and blacksmithing are limited quite strongly by their objects: only a small number of blacksmiths can stand at the forge, the metal affords a relatively small number of configurations, and customers might be able to give specifications but cannot work the metal themselves. In contrast, knowledge work involves specialists interacting heavily with each other and with textual resources as they transform knowledge at multiple points. Knowledge work objects are less material (knowledge represented in various texts as opposed to fields, game, or iron). They are more multiperspectival, worked by "multi-professional team[s]" (Engeström, 1992, p. 25) that share the objects and transform them according to their diverse fields of expertise, engaging in frequent boundary crossing to communicate their field-based assessments. Participants in such multiperspectival activities may not even agree on what the object is (The defendant or the case file? The patient or the patient's records?).

Furthermore, this increase in knowledge work is not restricted to certain sectors. Increasingly, literate work tends to be layered onto craftwork (cf. Karlsson, 2009) and agricultural work. Farmers who labor to transform their fields now use a GPS. Given these changes in work activity, Engeström's CHAT researchers are increasingly concerned with addressing knowledge work. In the past twenty years, various collections focusing on CHAT and Engeström's CHAT have addressed how it must adapt to discussing knowledge work (Daniels et al., 2010; Sannino et al., 2009; Sawchuk et al., 2006), as have various monographs (Engeström, 2008; Guile, 2010; Kaptelinin & Nardi, 2006; Spinuzzi, 2008, 2015a). Specifically, much knowledge work has become postbureaucratic, resulting in changes to how we organize and conceptualize work.

If work and work organization are changing, if the object of activity is primarily representational and thus more abstract, transformable, and circulable, then we might expect the concept of the object to develop in response. And so it has.

THEORY-MOVEMENT 1: KNOWLEDGE IS THE OBJECT; OBJECTS BECOME KNOWLEDGE

In method-movement 1, we saw that the activity system forms around an object, defined as the raw material or problem space (see figure 1.2). Knowledge work takes knowledge as its object. Knowledge, of course, is still instantiated in materials: a printed report, a database, a graph, a case file. But as knowledge, it can be represented in various materials, in redundant materials, even in an ecology of interrelated genres (cf. Ding, 2008). It can be copied, transformed, and fractured freely. When doctors "work" on a patient, they deal with a unique individual, but when they "work" on an article, they transform a printout that they can photocopy or a file that they can email. If they lose the patient, that's it; but if they lose the patient's records, they can usually get an archived copy.

THEORY-MOVEMENT 2: KNOWLEDGE IS A REPRESENTATION(S) OF ANOTHER OBJECT(S)

In method-movement 2, in chapter 6, we saw that the minimal model for CHAT-as-design-approach was two activity systems sharing the same object (figure 1.3f). But in Engeström's early examples, that object was doubly manifested. As we can infer from Engeström's (1990, 1992, 1999a) early studies, a contradiction exists between knowledge and that which it represents: the records versus the patient, the case file versus the defendant. The material object and its representation occupy the same space, often interfering with each other: doctors are distracted from the patients by their records and draw back to them when the patients go off script, while judges must repair the ruptures when defendants do not conform to the representations of their cases.

Of course, the patients and defendants are also objects in their own activities—with different motives. Indeed, in this theory-movement, objects are separated from motives, as Kaptelinin and Nardi (2006) noted (p. 157; cf. Kaptelinin, 2005; Nardi, 2005). The object expands to encompass multiple motives and to therefore become the nexus of two literate activities. It does this partly by encompassing representations of itself.

THEORY-MOVEMENT 3: KNOWLEDGE IS TRANSFORMED IN MULTIPLE ACTIVITIES; OBJECTS BECOME POLYCONTEXTUAL

In method-movement 3, we saw that CHAT spread outward to examine activity networks, first as atomistic, chained activities and then as overlapping, polycontextual activities (see figure 1.3g). Through these networks, representational objects circulate and are transformed by different specialists. Hospital staff may have trouble convincing a medical patient to move from one office to another, visiting different specialists, clerical staff, and insurance companies, and they may find it nearly impossible to get a patient to tell all these people the same story. But the staff can represent the patient in easily circulated, standardized records (Opel & Hart-Davidson, 2019).

Those working in the Engeström's CHAT tradition have characterized this phenomenon of transportability in various ways. For instance, some researchers working in Engeström's CHAT have borrowed Star and Griesemer's (1989) notion of *boundary objects* that circulate among activities (e.g., Engeström et al., 1995; Furstenau, 2003; Ludvigsen et al., 2003; Russell, 2010; Tuomi-Gröhn et al., 2003; Weber, 2003). Others have connected boundary objects with the concept of genre (e.g., Russell, 2009; Spinuzzi, 2008, 2010). Importantly, embedding the knowledge object in multiple representations means that it might fragment into multiple competing objects that interact fractionally with different activities (see chapter 9).

But transportability does not simply mean traversing a chained activity network: multiple interpenetrating activities begin to collapse into each other as specialists from different fields work on the same knowledge objects, transforming them to meet the requirements of their different fields—a common work arrangement in knowledge work (Adler & Heckscher, 2007; Toffler, 1980, 1990). For instance, the doctor and the clerical staff cooperate on transforming patients' records, but one does this to heal, while the other does this to ensure that the office can be paid. (And both transform these knowledge objects to avoid legal liability. Engeström's doctors do not want to become Engeström's defendants.) To work effectively, these specialists must cross boundaries (Tuomi-Gröhn et al., 2003). They must learn, not just vertically (about their own profession), but horizontally (about others' work) (Engeström, 2001; Tuomi-Gröhn & Engeström, 2003). Thus, multiple field and disciplinary frames are brought to bear on the same knowledge objects (Ludvigsen et al., 2003). Such activities become polycontextual (Engeström et al., 1995), with different specialists applying different frames to shared objects. Yamazumi (2009) proposed "hybrid activity systems" to address such interactions among activity systems (p. 213). In 1990, we could perhaps say with Engeström that "contexts are activity systems" (1990, p. 79). Twenty years later, we had to agree with Russell (2010) that "the network is the context" (p. 354).

The more representational such objects become, the more transportable they are (cf. Boisot et al., 2011). Workplaces increasingly become systems of fragmented knowledge, spread across space, time, and disciplinary activities (Bruni et al., 2007). Activities become placeless (Nardi, 2007, 2010). Ruckriem (2009) argued that the global communication networks that enable so much of this boundary crossing also challenge CHAT, since they enable pure contingency in self-weaving communication networks (p. 91; cf. Lompscher, 2006). Kaptelinin and Nardi (2006) argued that "work itself is changing. Work is more distributed, more contingent, less stable. How do we understand social forms such as networks and virtual teams that partially replace standard organizational hierarchies? . . . Knowledge work usually involves multitasking and working with diverse groups and individuals" (p. 26). And they conceded that "the conceptual apparatus of activity theory currently does not provide an elaborated set of concepts for the analysis of multiple activities" (p. 256).

Hence, the object expands again. Whereas the object was once understood as the linchpin that anchored the activity system, we now find that objects are embedded in multiple activities simultaneously. To understand a given object with its multivalence and multiple motivations, then, we must examine the networks of multiple activities brought to bear on a necessarily abstract object that is increasingly fragmented and contested.

THEORY-MOVEMENT 4: CO-CONFIGURATION EMERGES AS A HISTORICALLY NEW FORM OF WORK

In method-movement 4, we had to dissect a complex statement: "Runaway objects are transformed via knotworking in substrates of mycorrhizae." This string of methodological concepts points us to corresponding theoretical changes in Engeström's CHAT: as we saw in chapter 8, Engeström (2008) argued that new objects (runaway), collaboration patterns (knotworking), and communities (mycorrhizae) have developed to address a historically new type of production. The objects become larger—that is, spatially and temporally expanded (Engeström et al., 2003)—and because the object is the principle that bounds the activity system, the activity system becomes larger as well. Such expanded objects—objects that are addressed in different aspects by different activities, objects such as climate change and poverty (Yamazumi 2020, pp. 2–3)—become *runaway* in that they "are poorly under anybody's control and have far-reaching, unexpected side effects." (Engeström, 2008, p. 227). They are addressed by communities that are "weakly bounded and heterogenous *mycorrhizae-like* formations which rely on stigmergy, negotiation and peer review as coordinating mechanisms" (Engeström et al., 2010, p. 8, my emphasis). And collaboration in these expanded activities involves *knotworking,* which is "characterized by a pulsating movement of tying, untying and retying together otherwise separate threads of activity. The tying and dissolution of a knot of collaborative work is not reducible to any specific individual or fixed organizational entity as the center of control" (Engeström et al., 1999, p. 346).

This is not to say that all objects are runaway, all communities are mycorrhizae, or all collaboration is knotworking. But runaway objects, mycorrhizae, and knotworking are all concepts that emerge to address larger-scoped cases within the bounds of Engeström's CHAT, particularly cases that emerge from new types of production that are less bound by time and space than traditional craftwork. Such cases include

- agencies and neighbors coordinating to capture a psychiatric patient (Engeström et al., 1995);
- police investigating financial crimes (Engeström et al., 2003);
- a high-tech manufacturing company developing a product (Engeström, 2007; see also Daniels et al., 2010).

In each example, people from newly interoperating activities cross boundaries to co-configure their shared object. They knotwork, coming together in teams that are not necessarily persistent but that use mycorrhizal substrates to anchor the relationships between their activities. And in each example, we get the "spatial and temporal expansion of the object" (Engeström et al., 2003), and with it, a broader unit of analysis.

As Engeström (2008) argued, "the nature of teams depends on the historical type of production in which they are implemented" (p. 190)—and teams are but one of multiple "fluid forms of organizing collaborative work" (p. 194). Fluidity is characteristic of co-configuration work. As we saw in the discussion of method-movement 4, knotworked actors are not consistently the same, mycorrhizae are loosely held together and rely on alliances across rapidly changing activities, and objects are runaway, expanding in acute crises and breakthroughs.

Because of this fluidity,

> it is difficult for practitioners to construct a connection between the goals of their ongoing actions and the more durable object/motive of the collective activity system. Objects resist and bite back: they seem to have lives of their own. But objects and motives are hard to articulate: they appear to be vague, fuzzy, multifaceted, amoeba-like, and often fragmented or contested. The paradox is that objects/motives give directionality, purpose, and meaning to the collective activity, yet they are frustratingly elusive. (p. 204)

The more activities are brought to bear on an object, the more fragmented (or fractional; cf. Law, 2002) and contested it becomes. No one masters a given activity, which is now understood to have a changing object and motive (Miettinen, 2008). Activities' boundaries and certainties dissolve (Edwards, 2009). Activities now change so rapidly, become so interrupted and stochastic, that developmental spirals become more like eddies (Spinuzzi, 2008).

THEORY-MOVEMENT 5: SOCIAL PRODUCTION EMERGES AS A HISTORICALLY NEW FORM OF WORK

Co-configuration, as intermeshed as it is, gives way to social production (see chapter 8). As Engeström (2016) states, in social production, "activities take the shape of expansive swarming and multidirectional pulsation, with emphasis on sideways transitions and boundary-crossing" (p. 76). Boundaries across activities fade away (Engeström, 2009a). Engeström characterizes social and peer production as historically emergent work in which public goods are considered along with social problems and consequences (2008, p. 232); he ties it to Benkler's (2006) discussion of peer production enabled by new digital technologies.

Thus theory-movement 5 involved rediscovering or rehighlighting the political and thus answering critics who had characterized Engeström's CHAT as too apolitical (Yamazumi, 2020). In studying how cycles of expansive learning intermesh to address objects that are too big and diffuse to be addressed by individual activities, Engeström and Sannino also address the political dynamics involved in formulating, conceptualizing, valuing, and weighing composite and diffuse objects such as homelessness (see chapter 9). Yet, as objects of agreement, these objects of social production tend to be quite abstract. One can address an individual's homelessness by giving the individual a home. But to address ongoing homelessness at a national level, one must define it in policy, represent goals, and identify and measure outcomes across many affiliated activities.

Complicating this theory-movement is the fact that homelessness—proffered as an example of the diffuse objects resulting from social and peer production—is not particularly new, even if homelessness is on the rise. Neither, as Engeström notes, are other peer production examples such as birding (Engeström, 2009b; compare Prior & Schaffner, 2011; Smith & Prior, 2020) and skateboarding (Engeström, 2009b). We have long had diffuse activities cutting across public and private coalitions with differing interests. So is this theory-movement really the result of addressing historically emergent work, or is Engeström's CHAT instead encountering old problems as it expands its scope to address public policy issues (cf. chapter 9)? Is the problem new, or is it *newly visible* because Engeström's CHAT has broadened the unit of analysis?

SUMMARY OF THE THEORETICAL PROBLEM: SPATIAL AND TEMPORAL DIFFUSION OF THE OBJECT

This theoretical problem involves a second feedback loop. Engeström's CHAT brings ideas about individual and collective human development to the question of knowledge work; yet knowledge work's fluidity challenges this cyclical, developmental account of work, perhaps because fluidity is more visible to researchers in knowledge work (in which the researchers participate) than in agricultural and craft work (in which they generally do not). The problem is not just that we have trouble bounding the case, but that the cases are actually growing more interconnected and multidimensional, involving more representations and more interconnected representations. To make sense of them, Engeström's CHAT must account for multiple perspectives, polycontextuality, and emergence by again expanding the object.

THE FEEDBACK LOOPS EXPAND THE OBJECT: THE CENTER DOES NOT HOLD

Now we have at least two feedback loops: one methodological, in which the object expands to encompass the continually receding horizon of context, and one theoretical, in which the object expands to address the deepening multidimensionality of activities and the multiplying perspectives on any

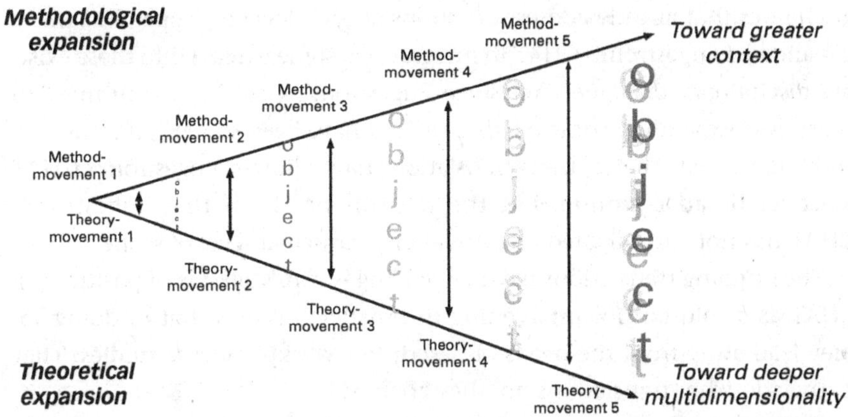

FIGURE 10.1
The theoretical and methodological expansion of the object

given object. See my own modest triangle diagram (figure 10.1). The methodological and theoretical expansions—associated with major translations we discussed in part II—each stretch the object, making it more abstract, more broadly contextual, and more multidimensional. In this expansion, each progressive version of the object defines an activity that is less bounded and coherent. *The object becomes progressively less explanatory*. As we'll see in chapter 11, Witte (2005; cf. Bracewell & Witte, 2003) arguably had a point when he complained that when an object is demonstrably used by more than one activity, it cannot serve to define an activity. Toward the end of the progression, multiple cycles of expansive learning address runaway objects such as global warming and homelessness—objects that are so diffuse as to defy a consistent definition.

The consequences of this methodological and theoretical expansion are significant. The story of Engeström's CHAT is the story of a series of translations, made in part to address more expansive objects, orienting to different disciplines. These translations result in ever-expanding *context*: the methods have to address greater spatial, temporal, and social ground, and the data look more like sociological data. They also result in ever-deepening *multidimensionality*: the theoretical apparatus has to address the additional perspectives and dialogue that become evident once the spatial, temporal, and social expansion in data occurs. Hence, Engeström's CHAT has steadily introduced new methodological and theoretical tools, largely in dialogue with existing disciplines that already address questions at each level of scope. This move has allowed Engeström's CHAT to integrate (to some extent) into these existing disciplinary dialogues. Yet since Engeström's CHAT has continued to chase ever-expanding contexts, these efforts have been somewhat transitory and fragmented. That is, these translations have allowed Engeström's CHAT to cover broader ground, but the tradeoff has been that Engeström's CHAT has not consolidated its gains at any particular level of scope.

The ongoing translations have served Engeström's agenda of positioning CHAT as a solution for intervening in broader contexts. But in doing so, they lead away from the needs of descriptive workplace case studies. That is, they do what translations do: they "betray."

In the next chapter, we take a closer look at how this expansion in scope has affected an often-invoked, but rarely discussed, term: the unit of analysis.

11 THE SCOPE OF THE UNIT OF ANALYSIS

With the previous chapters in mind, let's return to the two triangles we discussed in chapter 1. These diagrams are not just emblems, nor are they just helpful heuristics for describing activity to participants (although they have been used in both these ways). They additionally represent cultural-historical units of analysis. And examining them side by side helps us to understand how Engeström's CHAT has been translated to cover different levels of scope.

But the term *unit of analysis* is contested. Scholars use the term often but define it more rarely, and those definitions often do not match. Why? In this chapter, we examine the interpretive conflicts over the term, which are again related to CHAT's translations—specifically, the translations in the scope of the expanding object, described in chapter 10. We consider how these translations of the term *unit of analysis* imply different scopes and suggest different boundaries in workplace research.

To understand units of analysis, let's take one more detour: a detour to examine the most high-profile failure of the Vygotsky Circle.

"Uzbeks have no illusions!" So read a legendary, and perhaps mythical, telegram that Luria sent from Central Asia to Vygotsky in the early 1930s (Lamdan & Yasnitsky, 2016, p. 175; Homskaya, 2013, p. 26). Vygotsky and Luria had codeveloped the research design for this cross-cultural investigation, in which they hoped to demonstrate the social origins of psychological processes. The investigation was centered in Uzbekistan and Kyrgyzstan. In those territories, the population was largely illiterate. Soviet efforts sought to educate them, make them literate, and restructure their economy along Soviet principles. In 1931 and 1932, Luria conducted two expeditions, the results of which appear in his 1976 book *Cognitive Development: Its Cultural and Social Foundations*. There, he argued that Soviet-imposed literacy provided new cultural tools to Uzbeks. Such tools were seen as advancing the development of people who took them up: as Vygotsky and Luria argued in "Tool and Symbol in Child Development" (1930/1994), when an outward sign operation is transferred into a new intrapsychological layer, the new psychological system is "incomparably superior in content" (p. 155). In their Uzbek expedition, Luria and Vygotsky sought evidence that literacy and Soviet ideology indeed gave Uzbeks superior new capabilities. As Luria claimed in his retrospective preface,

> The masses [of Uzbekistan] had lived for centuries in economic stagnation and illiteracy, their development hindered among other things by the religion of Islam. Only the radical restructuring of the economy, the rapid elimination of illiteracy, and the removal of the Moslem influence could achieve, over and above the expansion in world view, a genuine revolution in cognitive activity.
>
> Our data indicate the decisive change that can occur in going from graphic and functional—concrete and practical—methods of thinking to much more theoretical and abstract modes brought about by fundamental changes in social conditions, in this instance by the socialist transformation of an entire culture. Thus the experimental observations shed light on one aspect of human cognitive activity that has received little scientific study but that corroborates the dialectics of social development. (Luria, 1976, p. vi; see Scribner & Cole, 1981 for a research project inspired by Luria's but with greater cultural sensitivity)

Luria's data were composed of interactions with the local populations, in which, among other things, Luria asked participants to examine figures (yes, including triangles) of well-known optical illusions. "Our goal was to determine whether the laws of perception described by the Gestalt psychologists were the same for subjects raised under different socioeconomic systems" (p. 31). As he presented these subjects with well-known optical illusions (p. 42), he found that the higher the subjects' level of education, the more likely they were to see the optical illusions (p. 43). He concluded, "Thus, the data clearly show that *optical illusions are linked to complex psychological processes that vary in accordance with socio-historical development*" (p. 43, his emphasis). Vygotsky was excited about these results, which seemed to validate his own ideas. He saw them as a breakthrough.

Yet the Central Asia expeditions turned out to be a disaster.

For the second (1932) expedition, Luria asked a gestalt psychologist, Kurt Koffka,[1] to witness these interactions. Just as Louis Pasteur arranged a demonstration of his vaccine in front of experts (Latour, 1988), Luria arranged to replicate experiments that had been successful in 1931, confident that he could demonstrate that the optical illusions that gestaltists had considered universal were actually rooted in social development. Yet, although the interactions went the way Luria expected, Koffka disagreed with Luria about what they meant. When Luria published his results in the *Journal of Genetic Psychology*, the editors asked Koffka to add his own views. About Luria's results, Koffka wrote,

> The exceptions, which were very rare in this expedition, but had been much more frequent in the first, are easily explained by the attitude of the testees towards the experimenter. Naive, social subjects who treated the experimenter on a footing of equality and did not regard the experiment as a test of their ability had the illusion without exception. *Only when the subjects were suspicious*, staring a long time at the patterns before making their judgments, the illusions failed to appear with some though by no means with all patterns, in accordance with well-known facts. . . . Thus the opposite report of previous investigators can be explained, according to which Uzbeks, not reached by modern culture, cannot perceive perspective. (in Luria 1934, p. 257, my emphasis)

Why would the Uzbeks be suspicious? In a letter to a former student, Koffka describes the expedition as a large convoy showing up unannounced in rural Uzbekistan villages—with presidents of Uzbekistan and Kyrgyzstan and the head of the secret police in tow (Harrower 1984, p. 150). With this situation in mind, we can see some of the Uzbek results as showing, not universal psychological results, but political caution and perhaps resistance. For instance, in one case, Luria presents this syllogism: "In the far north, where there is snow, all bears are white. Novaya Zemlya is in the far north, and there is always snow there. What colors are the bears there?" (Luria, 1976, p. 78). One Uzbek refused to give a straight answer, and when pressed, he responded: "Well, it's like this; our tsar isn't like yours, and yours isn't like ours. Your words can only be answered by someone who was there, and if a person wasn't there, he can't say anything on the basis of your words" (p. 79). Luria interprets this remark as an inability to understand syllogisms, but Matusov and St. Julien charge, "The elderly print-illiterate, unschooled Uzbek seemed to have more political awareness and courage than his interrogator when he directly referred to the political regime under which the syllogisms operate" (2004, p. 233). In contrast, the semiliterate and literate participants arguably answered in the way Luria expected, not because they had formulated a higher concept, but because they were schooled to respond with a particular performance (cf. Tulviste, 1991); they had accepted the power relations implicit in schooling (Bazerman, 2022; Matusov & St. Julien, 2004). Hannah Proctor, in her assessment of Luria's expedition, adds that Luria provides direct quotes from the Uzbek participants that contradict his interpretations and suggest different meanings; she notes Luria's apparent ignorance of the political pressures that might affect the Uzbeks' answers (Proctor, 2020). As Latour charges, Luria could have "let the story go on and added the context" (1987, p. 197), as Cole and Scribner did in their replication of this study in Liberia (Cole & Scribner, 1974).

Luria was soon on the receiving end of political pressures. Under the auspices of the Moscow Inspection Commission of the Workers and Farmers Inspectorate (Van der Veer, 2000), P. Razmyslov (1934/2000) wrote a report critiquing the Central Asia expeditions and the earlier work of the Vygotsky Circle. Razmyslov argued that Vygotsky and Luria could not look for primitive people's thinking in modern Uzbeks (p. 51). They had sought child-level behavior in these Uzbek adults rather than chronicling their growth (p. 51). Vygotsky's oral report of the expedition "was aimed at demonstrating the presence of primitive thought in all previously oppressed nationalities." Instead of showing how, in Uzbek, "the new [human] is being created and a communist consciousness is taking shape," Vygotsky tried to show that these peoples could not generalize (p. 52). Razmyslov drew several examples from Luria's investigations (pp. 52–54) and argued that Luria "literally tormented the respondents" (p. 53—perhaps a fair assessment). Luria and Vygotsky's cross-cultural research had been subjected to a trial and had failed that trial.

In response to this failure, Luria tried an unusual tactic: rather than translating his research, he simply outlived his critics. Luria put his book manuscript in a drawer for forty years, finally publishing it *with little alteration* in 1974. Koffka could no longer protest, as the book was published thirty-three years after his death.

This story illustrates the Vygotsky Circle's failure to interest the international psychological community, the Stalinist political apparatus, and the peoples of Central Asia. In examining how the Uzbeks perceived optical illusions, Luria and Vygotsky harbored their own illusions. And these illusions were due to Luria's failure to—as Latour put it—let the story go on long enough and to add more context. But what Latour presents as a simple task is actually a very difficult one. How does one select the appropriate unit of analysis?

WHAT ARE UNITS OF ANALYSIS?

In contemporary social science research, a *unit of analysis* is typically considered to be the entity that the researcher wants to study and what they want to learn about it. For instance, in the *SAGE Encyclopedia of Social Science Research Methods*, Long says that the unit of analysis is "the subject (the who or what) of study about which an analyst may generalize" (2004, p. 1158). That entity is generally determined by the research question and analysis. And in social science research, especially research studying broader phenomena such as organizations and geographic units, the unit of analysis is often a genuinely analytical construct that might not be studied directly.

What can be studied directly is the *unit of observation*, i.e., what the researcher actually observes, measures, or collects in order to learn something about the unit of analysis. Long differentiates them in this way:

> For example, political scientists may aggregate survey data at the country level by estimating the mean score on an item measuring political participation, assigning that same score to the country to identify how patterns of participation differ crossnationally. In this case, the unit of observation is the individual survey respondent, but the unit of analysis is the country. Likewise, achievements of students are often aggregated to allow schools to be compared. It is the country and the school about which the analyst generalizes, not the citizens or students. (p. 1158)

Long adds, "In the social sciences. . . . Researchers frequently examine the behavior of units that are unobserved" (p. 1158).

This broad understanding of units of analysis and units of observation must obviously be operationalized differently in different domains, disciplines, fields, and individual studies. For instance, in educational psychology, Säljö similarly defines the unit of analysis as

> the choice of a conceptualization of a phenomenon that corresponds to a theoretical perspective or framework. And every phenomenon lends itself to many conceptualizations. For instance, if one studies the traffic flow in urban areas, one attends to vehicles as parts of that system; how many they are, how much space they occupy, how fast they travel during peak traffic hours, and so on. There is no need for that particular inquiry to analyze the functionality of the electronic

> ignition system or the details of the tires. The interesting unit is the vehicle as a functional element of a traffic system. (2009, p. 206)

Since a phenomenon may have many conceptualizations, Säljö argues, the unit of analysis is related to the researcher's theoretical commitments as well as the phenomenon the researcher is trying to study. Thus, units of analysis cannot be disciplinarily invariant: some units are more suitable than others in a given study, due to disciplinary differences and relationships.

In their textbook on research in education, Cohen et al. (2000) describe units of analysis as categories used to code qualitative data: "Each unit of analysis (category—conceptual, actual, classification element, cluster, issue) should be as discrete as possible whilst retaining fidelity to the integrity of the whole, i.e. that each unit must be a fair rather than a distorted representation of the context and other data" (p. 148).

Also discussing qualitative content analysis of texts, Roller & Lavrakas (2015) caution,

> If a weak unit of analysis is chosen, one of two damaging outcomes will result. First, if the unit chosen is too precise (i.e., at too much of a micro-level than what is actually needed), the researcher will set in motion a content analysis study that may miss important contextual information and may require more time and cost than if a broader unit of analysis had been chosen. An example of a too-precise unit of analysis might be small elements of content such as individual words. In turn, if the unit chosen is too imprecise (i.e., at a very high macro-level), important connections and contextual meanings in the content at smaller (individual) units may be missed, leading to erroneous categorization and interpretation of data in the subsequent phase of the content analysis process. (p. 262)

Sociologists commonly use units of analysis such as the individual, group, organization or institution, cultural artifacts, social interactions, and geographic units. Again, these are selected based on the research question, and they are related to, but not always the same as, units of observation. Similarly, ethnographers commonly use micro, organizational, and macro units of analysis, although these sometimes rely on the same units of observation (such as interactions) (Jerolmack & Khan, 2017).

We'll return to contemporary qualitative research perspectives on the unit of analysis in a moment. Here, we need to note that in Vygotsky's day, the term "unit of analysis" was under dispute in the social sciences, meaning that Vygotsky and his contemporaries did not use the term in exactly the same way. This difference causes slippages, especially when contemporary

CHAT researchers use Vygotsky's concept of the unit of analysis to guide contemporary research.

THE UNIT OF ANALYSIS ACCORDING TO VYGOTSKY

In *The History of the Development of Higher Mental Functions* (1931/1997c), Vygotsky bases his discussion of units of analysis on Mikhail Basov, who in turn based his argument on gestalt psychologist Wolfgang Köhler. Using the analogy of the water molecule, Vygotsky argues that the unit of analysis should be the "real, objective element of which [a] process consists," just as the molecule H_2O is the smallest particle that retains the characteristics of water (pp. 66–67). Breaking the molecule down into hydrogen and oxygen atoms would not be useful for an analysis of water, since these individual atoms yield very different qualities. The task of analysis in psychology, he says, is "not to break down the psychological whole into parts or even into pieces, but to isolate certain traits and instances in each psychological whole that have retained the pre-eminent importance of the whole" (p. 67). These elements should not be isolated: instead, we should explain their connections and relations (p. 67).

Vygotsky lays out three principles for analysis in a genetic, developmental psychology: it must

- Analyze processes, not things (Vygotsky, 1997c, p. 68);
- Provide an explanatory study of the phenomenon, not a descriptive one (p. 68);
- Take account of "behavioral fossils" such as "automatic or mechanical mental processes" that have been developed and no longer reflect a current process (p. 71).

Vygotsky wrote this discussion during his instrumentalist phase, in which he mainly focused on what Wertsch (2007) termed explicit mediation, in which people use physical artifacts to control their behavior (see chapter 3). Consequently, Vygotsky took mediated action as his unit of analysis, examining how children used external signs to mediate memory, attention, and so forth.

As Vygotsky transitioned into his holistic phase, he became more interested in what Wertsch terms implicit mediation, in which individuals instead use internalized representations such as inner speech. At this point, Vygotsky turned to a second unit of analysis, which has been translated as "word meaning" (Vygotsky, 1986, p. 5) or "meaning through the sign operation" (Mahn, 2010, p. 315). His argument for analyzing by units rather than elements is still the same, as is the analogy of the water molecule (Vygotsky, 1986, pp. 4–5). But the phenomenon is different: "What is the unit of verbal thought that is further unanalyzable and yet retains the properties of the whole? We believe that such a unit can be found in the internal aspect of the word, in *word meaning*" (p. 5).

Vygotskians in the West have often taken Vygotsky's writings on units of analysis as the gold standard for understanding and evaluating them. But Vygotsky was not perfect—even if he did style himself as Moses (Zavershneva, 2016b, p. 126). In fact, Valsiner claims that in quoting Basov on units of analysis, *Vygotsky actually got the distinction backward.* Following Köhler, Basov "distinguished two kinds of 'structural elements' that can be observed in physical or psychological phenomena," which Basov term "real, objective elements" and "differential elements" (Valsiner, 1988, p. 174). Using the same analogy of water, Basov argues that water's elements, H_2 and O, are objective, real elements, while the particle of water (H_2O) is the differential element. As Valsiner characterizes it, Basov argues that the "real, objective" elements—the "qualitatively different sub-components" that make up a "qualitatively new structure"—should be the focus of the analysis, since their "*relationship* gives rise to new quality" (Valsiner, 1988, p. 175). Basov (and Köhler) preferred to use verifiable units and explore their relationships. Vygotsky misinterpreted this passage as endorsing the search for a "minimal whole" (p. 176). Valsiner argues that while Basov's approach "deals simultaneously with analysis (distinction of components in the whole) and synthesis (describing the nature of the connection of the components that create the super-componential new quality of the whole)" (pp. 176–177), Vygotsky's approach "analyzes more complex into less complex wholes, calling the latter units, without proceeding to describe how these units function" (p. 177).

Vygotsky's approach allowed him to focus on a holistic unit—but we can imagine how this focus limits a genetic explanation of a given phenomenon. It's hard to understand water *genetically* without understanding how the water molecule comes from the combination of hydrogen and oxygen (through burning hydrogen) or how those components can be split at some point (through electrolysis). In fact, when Vygotsky examined how word meaning arises genetically from the dialectical interpenetration of two independent elements, thought and word (Vygotsky, 1986, p. 212), he analyzed how those preexisting elements *formed* relations.

Indeed, as Zinchenko (1985, p. 99) acknowledges, Vygotsky was not always careful about his units of analysis. Despite his reputation as a *methodologist*, his actual *methods* often failed to yield durable results. We've already seen the example of the Central Asia expedition, in which the unit of analysis (mediated action, examined through discussions of optical illusions and syllogisms) was too restricted to understand the totality (Uzbek suspicion of, and resistance to, Soviet hegemony). Another example is that of Leontiev's small memory study (detailed in chapter 4), a thin reed on which Vygotsky and Luria staked a great deal (Vygotsky & Luria, 1930/1993, 1930/1994) without replicating or otherwise testing it.

One other thing to note about Vygotsky's understanding of the unit of analysis is that he did not distinguish it from the unit of observation. This is why the unit of analysis had to be "real" and "objective." We will return to this point in a moment.

THE UNIT OF ANALYSIS ACCORDING TO ZINCHENKO

Writing in 1985, Zinchenko reviews Vygotsky's writings about units of analysis and attempts to harmonize them with Leontiev's activity-theoretical school. Based on his readings, Zinchenko offers the following requirements for a Vygotskian unit of psychological analysis:

- "A unit must not be a diffuse, syncretic whole constructed of elements" but, rather, "an integrated psychological structure" (Zinchenko, 1985, p. 97).
- "The unit must maintain the characteristics of the unified whole, though internal contradictions and oppositions may exist" (p. 97).
- "These units, though preserving the structural characteristics of the whole, must be capable of development, including self-development" (p. 97)
- The unit must be "a unified system that cannot be further decomposed" (p. 98). This conclusion follows from the "structural integrity of the analytic unit, its heterogeneity (i.e., the presence of its contradictory origins), and its emerging properties," which imply that "the unit must be a *living* part of the whole" (p. 98).

- It must "begin with a taxonomic approach to the units of psychological analysis" (p. 98; Zinchenko credits Vygotsky with the implication, but credits Leontiev with its explicit elaboration).
- An analysis into units allows one to combine the analysis' "results with the synthetic of the properties of any complex unity" (p. 98).
- "The units of analysis" must "permit the investigation of the relationship between a specific psychological function (or process) and the entire life of consciousness and to the various critical functions of consciousness" (pp. 98–99).

As Granott notes, Zinchenko also "suggested another requirement: The unit has to be genetically primary. In other words, the unit has to be traced to the origination of mental processes" (1998, p. 45).

This is a long list indeed! In consolidating Vygotsky's pronouncements on units of analysis, Zinchenko perhaps translates them a bit, reading them through Leontiev's activity theory. Each item on the list progressively tightens the criteria for "units for the analysis of mind," applied specifically to psychological phenomena. That last point is critical to keep in mind: Zinchenko's criteria for the unit of analysis are so precise because the application of CHAT was so narrow in 1985, the year he wrote this chapter. But as we saw in chapter 6, two years later, CHAT was applied to much broader questions.

THE UNIT OF ANALYSIS ACCORDING TO ENGESTRÖM

In the introduction to the second edition of *Learning by Expanding*, Engeström summarizes one theme of his 1987 book: "the object-oriented and artifact-mediated collective activity system is the prime unit of analysis in cultural-historical studies of human conduct" (Engeström, 1987/2014, p. xvi). Elsewhere (Engeström, 2001, 2011; Engeström & Sannino, 2021), Engeström and colleagues characterize the "prime unit of analysis" of activity theory as evolving from mediated action to activity systems to activity networks. As we've seen, it is represented by his triangular diagram (figure 1.2).

Engeström uses this unit of analysis in his extensive qualitative studies, both in descriptive research and in his interventionist change laboratories. He does not explore the concept of unit of analysis in much detail, but he explains,

> The unit of analysis is *usually a conceptual idea strictly for the researchers*. In formative interventions, *we turn the unit of analysis into an external auxiliary means*, a mediating

> conceptual tool, for both the participants and the researchers. The triangular models of activity are typically presented and explained to the participants at an early phase of the intervention. Thereafter, they are repeatedly used to analyze and redesign various aspects of the participants' activity. (Engeström, 2011, p. 608, my emphasis)

And he adds,

> For formative interventions, the key implication of activity system as unit of analysis is that interventions need to be embedded and contextualized in the participants' meaningful life activity. An intervention that limits itself to the transformation of actions and ignores the motivational dynamics stemming from the object of the activity may be technically effective in the short run but is unlikely to have durable formative influence in the long run. (p. 608)

That is, he (like Bødker in her qualitative participatory design work) understood the activity system as a key way to *communicate with research participants*, who are understood as stakeholders—and he understood this shared unit of analysis as critical for changing, not just analyzing, a given activity (see chapter 6). This unit of analysis is centered on the shared object, which defines the activity under scrutiny; the boundary of the activity system is defined by the question of which subjects, community, and mediators engage in transforming it.

Thus, in Engeström's CHAT, activity systems really are *analytical constructs*, not the "real, objective" elements of Basov or Vygotsky (Valsiner 1988, p. 175). They necessarily rest on an analytical choice of which object to study, since the object provides the boundary of the unit. This shift to analytical constructs is itself a significant, hidden translation, and it is indexed to the shift from psychological phenomena to sociological ones. Engeström's analyses are oriented not to the mediated action of an individual in society or to the internal structure of the sign as manifested in word meaning; rather, they are oriented to labor activity at the scale of group or organizational collaboration among people of equal standing. Although this translation takes Engeström farther away from Basov, Vygotsky, and Zinchenko, it takes him closer to contemporary qualitative empirical understandings of the unit of analysis as opposed to a collapsed unit for both observation and analysis.

To understand this translation, let's take the question of medical work, which Engeström has studied throughout his career. We can distinguish several potential objects in medical work, including a *diagnosis*, which is typically conducted between a doctor and patient, sometimes with the help of other staff; *chronic illness*, which involves "a number of physicians, each viewing the patient from the perspective of his or her own specialty" (Engeström, 2018, p. 25); and a *treatment plan*, which can involve not just different specialists but also different institutions (hospitals, clinics, support networks).

These objects do not exclude each other. A diagnosis or a limited procedure can be conducted by a specialist, but it can be part of a larger phenomenon of chronic illness, which can in turn be addressed by a treatment plan. *Selecting an object involves determining the scale at which the researcher is working.* And as we've seen in this and previous chapters, larger scale involves a larger number of perspectives, a broader set of data, a different set of qualitative methods, and a lengthier intervention. This aspect of pragmatic intervention is critical for understanding what Engeström means by unit of analysis.

Bounding the unit of analysis in this way—via the object—serves the goals of an Engeström's CHAT analysis by segmenting and analyzing the work of a given group, organization, or institution. Yet we are engaged in several objects (and the activities associated with them) simultaneously (e.g., Russell, 1997a), and different analytical goals suggest different units of analysis. For instance, as Engeström expanded his empirical scope in the 1990s and 2000s, Susanne Bødker continued focusing on examining detailed streams of work in which individuals encountered different objects and used different mediators to address them (e.g., Bødker & Andersen, 2005; see also Arvola, 2004, 2014). These different boundaries reflect different interventionist foci: Engeström has increasingly tended to focus on interventions via broad coordinative genres such as care agreements, while Bødker and other participatory designers have tended to focus on intervening to redesign specific interface features, equipment, and spaces.

In defining the activity system as a bounded unit of analysis, Engeström opens the possibility for not just scaling units of analysis, but also comparing them in their connections. Engeström does this early on by proposing the notion of the network of activity, in which at least two activity systems interact (figure 1.3f), and potentially more can interact (figure 1.3g). In the case of two activity systems interacting, Engeström envisions two durable activities that take the same object but try to transform it in different ways; that is, the object is the same *objectively* (it constitutes the same raw material) but not *projectively* (the different activities project different desired futures on it). Consider a doctor and a patient during a medical consultation:

> The object moves from an initial state of unreflected, situationally given "raw material" (object 1; e.g., a specific patient entering a physician's office) to a collectively meaningful object constructed by the activity system (object 2, e.g., the patient constructed as a specimen of a biomedical disease category and thus as an instantiation of the general object of illness/health), and to a potentially shared or jointly constructed object (object 3; e.g., a collaboratively constructed understanding of the patient's life situation and care plan). (Engeström, 2001, p. 136)

In this move, Engeström identifies a unit of analysis (an activity system) with two instantiations, and he describes how to link them analytically. The result is what Bakhurst characterizes as modeling organizations, but what qualitative researchers might characterize as *linked case studies*.

As Witte (2005) argues, Engeström's triangle is not in itself a unit of analysis, and neither is Vygotsky's triangle. Both are simple, synchronic *representations* of the respective units of analysis (p. 142). Engeström's triangular diagram, in particular, has received much attention and criticism for the unit of analysis that it represents. Perhaps the most detailed critique comes from Andy Blunden.

THE UNIT OF ANALYSIS ACCORDING TO BLUNDEN

In his book *An Interdisciplinary Theory of Activity* (2010) and his edited collection *Collaborative Projects* (2014a), Blunden criticizes Engeström's activity system, which he finds to be inadequate as a unit of analysis for at least three reasons:

- It is not a concept, but rather a system (Blunden, 2010, p. 232).
- It attempts to include context, making this unit of analysis an open-ended totality due to infinite regress (p. 251).
- It repeats Leontiev's disconnection of motive and goal, which (Blunden points out) removed the ability to define the activity from its participants, instead resting that power in the hands of those who manage society (p. 210; cf. chapter 5).

Pointedly, he charges that "Engeström either does not understand what a unit of analysis is, or he holds that a unit of analysis is not required for Activity Theory" (p. 213).

In contrast, Blunden argues that the unit of analysis must be based on the germ cell. Working in the tradition of Hegel, Marx, and Engels, Blunden argues that the essential unit of analysis must be

1. A conception of a singular, individual thing that
2. Exhibits properties of a class of more developed phenomena and
3. Is itself an existent phenomenon (Blunden 2010, p. 190).

Thus, for Blunden, the essential unit of analysis is the "cell," which can "provide the building block for *conception* as well as *actuality*" (2010, p. 191). "A science," he argues, "can only base itself on something real and empirically given" captured as a concept (p. 190). That empirically real thing, he adds, must develop from a deep understanding of the relationships under analysis. For instance, in education, "when we grasp the situation as a predicament, a trap from which the child must emancipate herself (Borozhov 2004), then we have what is *both a concept* and *an existent reality*" (2010, p. 191; cf. Blunden, 2020).

He proffers a unit of analysis different from Engeström's activity system: the *collaborative project*, which is a joint proposed undertaking (Blunden, 2010, p. 256). Such a project has a concept, but each participant applies a personal meaning to it (p. 257)—that is, he corrects Leontiev's separation of motive from goal. This capacious unit of analysis, he argues, can be applied to phenomena as constrained as people's attributes and artifacts as well as more expansive collaborations (pp. 257–258): the projects can "nest" in each other (p. 262). Blunden sees this unit of analysis as being part of his aim to resolve "those features of CHAT which have prevented it from fulfilling its potential as an interdisciplinary approach to the human sciences in general" (2010, p. 3). He sometimes positions the collaborative project as conciliatory with Engeström's activity system (e.g., Blunden, 2014c, p. 25) and hopes to appropriate the empirical work that Engeström and others working in this vein have done (Blunden, 2014b, pp. 371–372).

Based on what Blunden says in *An Interdisciplinary Theory of Activity* (2010), he understands the collaborative project as being scalable in a way that the activity system is not: applying equally to artifacts, dyads, organizations, and larger and more complex empirical cases. Whereas Engeström has addressed the expanding object by chasing after larger contexts with the help of larger interconnected units, Blunden proposes to use the same unit at all scales. The diagnosis, treatment plan, chronic illness, and national health infrastructure could all be taken as collaborative projects, just as they are taken to be the objects of activity systems, but they are not linked—a single instance of the unit of analysis encompasses the entire field under investigation (in accordance with Blunden's requirement 1 above). Finally, like Vygotsky, Blunden does not distinguish between the unit of analysis and the unit of observation (in accordance with his requirement 3 above).

Theoretically, the solution is elegant. *Methodologically*, however, it is questionable. It's difficult to operationalize Blunden's unit of analysis for actual empirical studies: What data are collected? What are the units of observation? How are they related? How are units of analysis compared? If units of analysis can exist at different scales, what happens when one unit of analysis (a text containing a diagnosis) is compared with another (the national health system in which the diagnosis was produced)? Is it possible to link different instantiations of a unit of analysis to understand how they interact? (Based on Blunden's argument, I think the answer to this last question would be a resounding no.)

This operationalization problem is compounded by the fact that Blunden does not appear to do (or have a background in) empirical research. Yes, his edited collection *Collaborative Projects* does contain several empirical cases meant to operationalize the collaborative project as a unit of analysis for empirical research (Blunden, 2014c). But these cases do not actually accomplish this goal. We can see this in the collection itself. John-Steiner points out that these studies actually use different definitions of the collaborative project (2014, p. 114). Worthen notes that these research studies are told as narratives, rather than as research studies (2014, p. 357)—and as she does *not* point out, they are thus light on methodological description. Finally, in Blunden's own chapter under the section "Research Projects" (Blunden & Arnold, 2014), he and his coauthor provide a narrative recollection of an incident rather than a research study. They do not articulate a methodology, nor do they cite methodological literature applicable to their method, nor do they attend to basic questions of empirical social science research such as triangulation.

In sum, these studies do not appear to use the same concept of collaborative project as unit of analysis, and they do not demonstrate how that concept could be operationalized or integrated into empirical methodology.

Aside from the problems in operationalizing this unit of analysis within an empirical framework, the central issue is that Blunden's understanding of the germ cell brings us back to the problem for which Kozulin criticized Leontiev and the Kharkov Circle. The analytical principle is the phenomenon; the map is the territory. The phenomenon is understood as out there, independent of participants' perceptions, even though they participate in forming and sustaining it. In practice, this means that the researcher finds what they set out to find. In a Hegelian frame, this issue may not appear to be a problem. But it is recognized as a significant problem in basic qualitative research methodology. Recall Luria and Vygotsky's Uzbek research expeditions described earlier in this chapter, in which Luria set out to explore how the Uzbeks formed concepts, unaware of the naked power dynamics that influenced his participants' answers. Elsewhere, anthropologists conducted research *without* the backing of intimidating security services—and their participants felt much more free to resist, mislead, shut out, or ignore them (e.g., Bateson, 1958; Evans-Pritchard, 1940). Blunden criticizes Leontiev for taking the managerial view as a God's-eye view but gives us no way to avoid this misstep when identifying the essential shared concept that makes up a collaborative project—and this omission is significant indeed, since the history of anthropological and sociological research is replete with examples in which a concept is projected by the researcher but not shared by the researched (e.g., Fried, 1975).

Thus, Blunden's unit of analysis does not provide an empirical path for understanding the lived, sensuous experience of independent agents (but see Jornet & Damşa, 2021, for an attempt at reformulating it for these purposes). Those agents rarely see things in the way the researcher expects, and they rarely conform to the researcher's expectations, as experienced field researchers have ruefully discovered (e.g., Becker, 2008; Rabinow, 2007; Van Maanen, 2011; Whyte, 1984). In such situations, the phenomenon cannot be taken as self-evident, and the difference between a unit of analysis and a unit of observation becomes critical. For that reason, qualitative research methodologists do not use the term *unit of analysis* in the same way Blunden advocates, partly because Blunden's definition is difficult or impossible to operationalize in a qualitative research study—a real limitation for Blunden's approach, which aspires to become interdisciplinary yet disallows the translations that would be required in order to address the research questions and methodological wisdom that other disciplines offer. And tapping into that broader expertise is critical, since (as discussed in part II and summarized in chapter 10) Engeström's CHAT has scaled up to social systems, becoming a de facto organizational sociology that draws heavily on qualitative research methodology. Faced with different research questions, researchers cannot rely on a single, universal unit of analysis—they require different units tuned to answering different questions.

THE UNIT OF ANALYSIS ACCORDING TO MATUSOV

Eugene Matusov makes this point in a 2007 article reviewing the discourse surrounding units of analysis in sociocultural circles. After noting a proliferation of units of analysis, most of which have been covered in this chapter (word meaning, mediated action, activity system, interacting activity systems), he noted that "sociocultural scholars definitely preferred certain units of analysis not just for some of their empirical studies but, it seems, in general or even universally. When they move from one unit of analysis to another one as 'the unit', they seem to be saying that now they have discovered a better one" (2007, pp. 307–308).

Yet, he argues,

> I have learned that we, sociocultural scholars, should be careful in our search for the general or even the universal unit of analysis because the unit(s) of analysis as well as the analysis as a whole analysis is necessarily shaped by the purpose of the researcher and the material of the study. How can a certain unit of analysis be universally appropriate for any sociocultural research, disregarding its purpose, the object and subject of study, the targeted audience, expected practical and political consequences, and the researchers themselves?" (p. 308; cf. Säljö, 2009)

In the subsequent discussion, Matusov notes four things about the concept of the unit of analysis in this discourse:

- The term "unit of analysis" is used polysemously: both descriptively (that is, in describing a phenomenon) and prescriptively (that is, in analyzing it) (Matusov, 2007, pp. 309–310). It is sometimes treated as the map, sometimes as the territory.
- Vygotsky applied his units of analysis (word meaning, mediated action) to low-level phenomena. But Matusov argues that "a unit of analysis may not be appropriate for studying one phenomenon but may be appropriate for studying another. In contrast, contemporary scholars working within a sociocultural framework would apparently not admit the usefulness of studying the 'individual as unit of analysis' for whatever phenomenon" (2007, p. 310). Here, Matusov is not arguing for using the individual as a unit of analysis, but rather questioning why a single unit of analysis should be used for all phenomena.

- A universal, self-contained unit of analysis—with its inevitable partitioning—may not even be possible when studying human activity (p. 310). Matusov notes that "Gestalt psychologists and Vygotsky insisted on selecting and studying only a phenomenon that is self-contained and involves the unified smallest whole. . . . Their physical examples of homogeneous physical objects like water or air supported the idea of partitioning phenomena into units" (p. 310). This work drew an analogy with physical sciences, as Vygotsky's contemporary Lewin (1935) did. But such partitioning does not scale well.
- This partitioning comes from modeling the social sciences after natural sciences, treating participants with voices as "physical voiceless phenomena" (p. 310; cf. Matusov & St. Julien, 2004).

Unlike voiceless phenomena—and unlike participants that have faced power differentials in early CHAT studies, from schoolchildren to rehabilitating soldiers—participants in workplace studies are generally on a more or less even footing with the researchers. They enact meaning for themselves. This point was critical to the Western alteration of CHAT: Engeström and Bødker (chapter 6) both proposed deliberative processes by which participants could collectively make meaning. This move invited the participants to actively participate in interpreting and deliberating over their own work, not just to be studied. By drawing from Western qualitative research methods, Engeström and Bødker made the research a more open-ended process in which participants' interpretations and accounts were valued along with the researchers (although, arguably, the researchers' frameworks guided these interpretations). Engeström's triangle was used as a *language game* (cf. Ehn, 1989) to ground and elicit this participation. This use is central to his method but absent from Vygotsky's, Zinchenko's, and Blunden's.

With these four criticisms in mind, Matusov notes the issue of expansion discussed in chapter 10: in trying to develop a universal, holistic unit of analysis, sociocultural scholars find themselves "embracing bigger and bigger systems of practice, community, relations and culture" (2007, p. 323). Such an expanded unit of analysis is "more holistic but less manageable. Researchers simply cannot study everything and 'travel' unlimitedly over open networks of practices to trace their hybridity as well as multiple affiliations of their participants" (p. 324).

Furthermore, a universal unit of analysis in sociocultural research "does not depend on the researcher, his or her goals and foci," and "is not interpretative":

> Although it is true that different sociocultural scholars have different units of analysis, their discourse is a suspiciously modernist discourse searching for the universal truth about the true unit of analysis for any research rather than interpretative discourse about the particular contexts and the particular foci of particular research. It seems that a sociocultural approach betrays itself with its strong emphasis on context. (2007, p. 325)

Consequently, Matusov argues,

> the discourse on the unit of analysis in a sociocultural field has another related problem of being monologic and, thus, hegemonic. The sociocultural search for a uniting universal *It* as the unit of analysis is probably rooted in the monologism of Hegelian dialectics. . . . The issue here is of stratified power relationships between researcher and the people studied, which are actualized all the time in the official research accounts of published research. (2007, pp. 325–326)

Although writing three years before Blunden's 2010 book, Matusov neatly identified the problem with applying Blunden's proposed unit of analysis to empirical cases: "What is a universe for one person is not a universe for another" (2007, p. 326; cf. Escobar, 2017).

Matusov sums up: "I propose that units of analysis have to be always viewed as partial, incomplete and open . . . defined in part by the studied object, in part by the researcher's focus, in part by the audience of research and in part by the research participants (as distinct from the research object)." And, "A study should not just be a story about third persons for an academic community but also *a dialogue with people who participated in the study*. . . . Not only do study participants' voices have to be presented in the unit of analysis, but the analysis itself has to be *presented and heard by the study participants*" (p. 328, my emphasis).

Although Matusov does not use the language of translation, we can see parallels. Those in CHAT who have insisted on fidelity to the works of Vygotsky (and Hegel and Marx) do not want it to betray that heritage. But as we've seen, *translation entails betrayal*: to address phenomena beyond the individual mind in society, to perform interessement across new actors, the unit of analysis *must* be translated. This does not mean that Engeström's triangle represents a problem-free translation, obviously—but it does mean that the unit of analysis must be more flexible, more independent of Vygotsky's texts, than some CHAT theorists are willing for it to be.

Matusov's criticism of the unit of analysis may be an outlier in sociocultural studies, but it is in line with those of contemporary qualitative research methodologists. I'll turn to these next, since in its translation to the West, Engeström's CHAT has been applied in concert with Western qualitative research methodology to explore broader social systems (chapters 6–9 and this chapter).

THE UNIT OF ANALYSIS ACCORDING TO CONTEMPORARY QUALITATIVE RESEARCH METHODOLOGISTS

Earlier, I pointed out that Engeström characterized the unit of analysis as "usually a conceptual idea strictly for the researchers" (2011, p. 608). Which researchers? Engeström is not referring to Basov, Vygotsky, Leontiev, or Zinchenko here: his frame of reference is Western educational research. He cites contemporary Western journals such as *Research Policy*, *Educational Researcher*, *The Journal of the Learning Sciences*, and the *Review of Educational Research*. It is this group of researchers he is trying to interest. And it is in reference to such educational researchers that Säljö defined the unit of analysis as "the choice of a conceptualization of a phenomenon that corresponds to a theoretical perspective or framework" (Säljö, 2009, p. 206). Following Säljö, we must consider questions such as these when selecting a unit of analysis: What research question is the researcher trying to answer? What phenomenon are they trying to study, and how is it operationally defined? What unit of analysis is appropriate for the field in which the research is being conducted?

As we saw in part II, and reviewed in chapter 10, CHAT has been repeatedly translated into different fields, taking on phenomena that are progressively expanded in temporal, spatial, and social terms. Especially as it has been taken up in workplace studies, CHAT has been articulated in terms of interactionist design (chapter 6), case studies of workplace writing (chapter 7), and organizational sociology (chapter 8)—but not (yet) in terms of public policy (chapter 9).

So what is involved in these progressive translations? Let's return to the discussion of units of analysis that was being conducted in the 1920s. In an article referenced by Long in her encyclopedia article on units of analysis in social science, Fred Boucke (in 1923, the year before Vygotsky was hired by Kornilov) argued that "the units of social science are not things, but relations. They are relations between objects and human beings, or between individuals or social groups. Or they are relations between things brought to bear upon other units of psychic origin. In this sense, too, the units of social science are composite. They nearly always terminate in norms and interests of human beings" (Long, 2004, p. 446).

And "anything almost may be made a unit for measuring relations and finding constants of change" (p. 446). These "units are necessarily composite, and center in the ideas, interests, and efforts of human beings" (p. 443).

Long comments on Boucke's argument:

> Boucke (1923) notes two differences that distinguish social units from the units of analysis common in the natural sciences: First, the units of social science research are not measurable in a straightforward way or on a determinate scale. . . . Second, as evident in the examples presented above, units of analysis in the social sciences can usually be divided into various subunits: Communities may be divided into families or households, geographic neighborhoods, or individual members. Boucke concludes that units of analysis in social science are not "things" but the relationships or networks connecting the families, neighbors, or, at the most basic level, the interests of individuals. (Long, 2004, p. 1158)

Arguably, Boucke is too simple here (at about the same time he made his argument, the natural sciences themselves were rethinking measurability due to quantum mechanics), but his critique nevertheless applies to Vygotsky. While Vygotsky sought units of analysis that could be considered real and objective—units of analysis that coincided with units of observation—social scientists working at other levels of scope seek units of analysis that are necessarily analytical constructs representing composite relationships. These are generally separate from units of observation. When a social scientist studies a community, an organization, a geographical region, or a nation, they are not collecting a whole community as a piece of data; they are sampling data that can help them understand the internal relationships within that unit of analysis as an analytical construct.

We specifically see this understanding of units of analysis in contemporary qualitative research. Although qualitative research can certainly be conducted under a positivist paradigm, most contemporary qualitative researchers operate under postpositivist, interpretivist/constructivist, radical/critical, or poststructuralist paradigms (Guba & Lincoln, 1994). These paradigms emphasize inductively exploring and describing how people make meaning. Like Matusov, qualitative researchers in sociology, anthropology, and related fields (including writing studies) have generally been concerned with how to enter into the multivoiced, dialogical lives of their participants, balancing emic and etic understandings of what is under investigation—whether it is a biography, a bounded case study, or a culture (Creswell, 2006). That is, they are generally interested in developing a thick description (Geertz 1983) of a culture or a case, one in which readers can understand the different explanations, meanings, and interpretations of different participants. They do not seek to develop a unitary concept so much as to understand the different, fragmentary concepts at work.

For this reason, basic texts on qualitative research methodology take the unit of analysis specifically as *analysis*, not the phenomenon to be studied. The unit of analysis is recognized as an analytical decision driven by the research question, not an underlying phenomenon—a pragmatic decision rather than a description of an objective reality, a map rather than a territory. Yes, the unit of analysis "should . . . be discrete as possible whilst retaining fidelity to the integrity of the whole" (Cohen et al., 2000, p. 148), but that goal is achieved via an *analytical* process such as thematic coding.

In basic case study methodology, for instance, the case is a bounded, comparative unit. Bernard says that "in a case study, there is exactly one unit of analysis—the village, the school, the hospital, the organization" (2002, p. 39). Similarly, Yin argues that "as a general guide, your tentative definition of the unit of analysis (and therefore of the case) is related to the way you have defined your initial research questions" (2003, p. 23). Yin adds that for comparing cases with those in previous and subsequent research, "each case study and unit of analysis either should be similar to those previously studied by others or should innovate in clear, operationally defined ways" (p. 26). Creswell argues that

> case study research is a qualitative approach in which the investigator explores a real-life, contemporary bounded system (a *case*) or multiple bounded systems (cases) over time, through detailed, in-depth data collection involving *multiple sources of information* (e.g., observations, interviews, audiovisual material, and documents or reports), and reports a *case description* and *case themes*. The unit of analysis in the case study might be multiple cases (a *multisite* study) or a single case (a *within-site* study). (2006, p. 97)

Creswell cautions that "the key here is to define a case that can be bounded or described within certain parameters, such as a specific place or time" (p. 98). Merriam and Tisdell (2015) argue that "the unit of analysis, *not* the topic of investigation, characterizes a case study" and suggest either a bounded system or a specific instance such as a carefully selected individual (pp. 38–39).

In each of these instances, the unit of analysis is interpretive, depending on the research question and thus the goals and foci of the researcher (cf. Matusov, 2007, p. 325). In some cases, such as participatory action research and other forms of interventionist design research, the research question, goals, and foci are negotiated with the participants as well. Researchers recognize that they may not understand a phenomenon or its boundaries in the same way as participants, and participants may in turn not share these understandings with each other. Thus, the unit of analysis must be an analytical construct because when dealing with social phenomena, we do not necessarily share a single phenomenon. We share relatively coherent enactments.

One example of this understanding, grounded in actor-network theory (ANT), comes from Annemarie Mol, whose *The Body Multiple* investigates atherosclerosis. In her ethnography, she asks different specialists what atherosclerosis is—and she receives radically different answers, not just materially or methodologically, but also conceptually. Different people's versions of atherosclerosis are enacted differently (and Mol selects the term *enacted* carefully, to indicate the complex practices in which they are embedded). Usually, these multiple enactments of the disease cohere—that is, there's enough correspondence among them that people can be said to be talking about the same object, the same disease. But, interestingly, sometimes these enactments don't cohere. The research becomes "messy" (Law, 2004).

Another example, grounded in CHAT, is that of Jornet and Damşa (2021). They argue that in learning theory, sociocultural units of analysis have tended to become oversimplified and dichotomized in their distinction between the individual and the social. In response, they advocate an ecological approach to inquiry, which "naturalizes the process of arriving at concepts and categories, making it part and parcel of the way life activities unfold in actual life contexts" (pp. 3–4). This approach implies the following:

- "Ecological units denote evolving social wholes that are not pre-established but actually found in and through inquiry" (Jornet & Damşa, 2021, p. 4). That is, ecological units of analysis emerge inductively, since they study "*events, experiences and situations*, which are messy and holistic in nature"

(p. 4, their emphasis). Like Blunden, they draw on the language of the germ-cell to identify a core phenomenon that develops through concrete history.

- "To investigate and come to know an object involves relating to it both intellectually and affectively through concrete experience/activity" (p. 4). In inductively formulating ecological units, researchers develop both intellectual and affective experience with their emerging unit of analysis.
- "The validity and rigor of observations, from an ecological perspective, is determined by relevance to practice" (p. 4). Drawing on pragmatism, they argue that studies using ecological units are not focused on causal relationships but, rather, on how processes and products "relate to future opportunities for development" (p. 5).

Overall, in qualitative research methodology, actual lived experiences are understood as multifaceted, and participants interpret them (i.e., make meaning of them) for themselves. They often—and appropriately—resist the interpretations of the researchers involved, especially if these interpretations might result in changes to their actual working lives. As we saw in chapter 6, this fact has in part driven Westernized CHAT's shift to interventionist research: participants have a forum for collectively interpreting findings and deliberating about their implications. Yet they are guided in this interpretation and deliberation through the mediation of activity system diagrams, which function both as a *heuristic* (or dissection of context) and a *representation* of an underlying universal structure (in the mode of classical sociology).

With the above in mind, let's return to Witte's critique of the activity system as a unit of analysis. Witte argues that Engeström's model has a unit of analysis problem: if everything can be an activity, how does one bound and analyze a specific activity? (2005, p. 141). Engeström does not directly answer this question. But one way to do so is to *translate the concept of activity into case study terms,* which is the tack that we have generally taken in workplace writing studies (e.g., Spinuzzi, 2018, 2023b). That is, the researcher identifies a case by identifying an object—a material phenomenon that is both "objective" (material) and "projective" (intended for transformation). The object then functions as the boundary condition for the rest of the case: What actors are directly working to transform it? With what mediational means? Using what rules and division of labor? For the benefit of what imagined or named community? This boundary is not "predicated on some consideration that is external to the theory itself," as Witte (p. 140) argues, but on the object, which is well grounded in the theory. With this approach, we can investigate different scales by choosing appropriate objects: for instance, we could investigate a doctor-patient consultation as having the object of an individual patient's health, but we could also investigate a hospital unit as having the object of many patients' health and a national health care system as having the object of the cumulative health of a nation. These are different objects operating at different levels of scope, although they do notionally overlap, and they suggest differently scoped boundaries for their different cases. We can also draw on the vast literature on linked case studies, understanding activity networks as linked cases.

This interpretation, of course, is yet another translation. In translating the activity into case study terms, we can avail ourselves of established case study methodology, find further support, and identify suitable units of observation as well as data collection techniques that can yield them. We can also retain the basic categories of the activity system, not as elements but as a "composite," as "not things, but relations" (Boucke, 1923, p. 446). We can—as case study methodologists already do—study activities as developing. And we can find both methodological justification and operationalizable guidelines for addressing different levels of scale.

CONCLUSION: TRANSLATING THE UNIT OF ANALYSIS

Let's end the discussion by returning to Luria's ill-fated expedition of 1932. In that expedition, Luria sought to demonstrate that psychological phenomena that had broadly been considered *natural* were actually *cultural* and, specifically, represented the cultural gains yielded by education. He was confident that these results, which had already been evidenced in his 1931 expedition, would convince Koffka during the 1932 expedition. Yet he was so narrowly focused on the unit of observation that he could not see what was evident to Koffka: participants did not interpret these interactions the way he did. They were not willing participants[2] in either the study or the cultural transformation that had been foisted upon them. We can imagine what they might think of how Luria described the cultural context in the preface to his book: "economic stagnation and illiteracy" in which development had been "hindered . . . by the religion of Islam" (1976, p. v).

What Luria missed, and what Koffka observed, was outside of Luria's narrow unit of analysis: the deep suspicion and distrust among the least educated (and perhaps least compliant) participants. Setting out to study optical illusions, he was unwilling to admit that his own results might be illusory. Like Vygotsky, Leontiev, and Sannino, Luria interpreted resistance as a defect. Dialogue was all well and good, but it ultimately had to contribute to a dialectical synthesis.

Rather than translating his research approach to address the concerns of Koffka and Razmyslov, Luria instead avoided the dialogue with them—publishing after his critics were dead.

Luria's option is not available to CHAT if it is to continue to grow and develop. Between 1932 and now, CHAT has greatly expanded its scope, chasing after the context that Luria missed. But to do so, CHAT has had to reconsider its unit of analysis.

We've seen two basic paths in CHAT work:

- The first path is to *reject* translation, sticking with the requirements for a unit of analysis that Vygotsky set out in the 1930s, insisting that research in other disciplines becomes reconfigured in service to these requirements. But this path leads to the marginalization of CHAT. Vygotsky may have styled himself as Moses, but he was hardly infallible, and his research questions and concerns do not match those of the expanded scope to which Engeström's CHAT has attempted to apply itself. This path is a dead end.
- The second path is to *embrace* translation, seeking to understand units of analysis in other traditions and striking new settlements that make CHAT useful to and compatible with these traditions. Engeström's CHAT has followed this path to some degree, although with some obvious missteps, discussed in chapters 6–9. These settlements have held—to an extent.

But settlements are fragile. In the next chapter, we will review the current settlement of Engeström's CHAT and what it tells us about social science research.

12 THE DIFFUSION AND TRANSLATION OF CHAT

Recently, a senior scholar in CHAT (not Engeström) confidently told me that Vygotskian theory can "save the world." As you may have gathered, reader, I have some doubts about this assertion. Despite his best efforts, Vygotsky did not provide us with a totalizing framework suitable for all interdisciplinary research, much less for saving the world. None of the triangles we encountered in chapter 1 can serve as an enduring settlement. In fact, treating Vygotsky's contributions as if they are universally applicable and inviolable *guarantees* that others will not take them up: no betrayals, yes, but no translations either.

My interlocutor was concerned with saving the world. But as mentioned at the beginning of this book, I'm more concerned with the narrower questions of how people communicate, coordinate, and collaborate at work and how they might do these things in ways that improve their collective and individual lives. During a time of tremendous changes in work—in information technologies, organizational structures, societal expectations, and even in how we commonly define and conceptualize "work" and "workplaces"—even these narrower questions are difficult to answer. But the answers can concretely impact the quality of people's lives. It's with that project in mind that I originally took up CHAT, nearly thirty years ago, to shape my research agenda. It seemed custom-built for my purposes and those of many in writing studies. Taking it up allowed me to address my goals: over the intervening time, I have explored the work of software engineers, traffic engineers, telecommunications workers, search marketers, nonemployer firms, and early-stage technology entrepreneurs, charting the ecologies of genres they bring to bear on their work. Engeström's CHAT has provided the theoretical and methodological framework for those studies, giving me the concepts and vocabulary to share insights with these participants and think through their work with them. So I'm glad that I took up this framework.

But, as we've seen throughout this book, "taking up" a social science framework like CHAT is not like taking up a single instantiation of a tool, like taking a hammer out of a toolbox only to return it at the end of the workday. Social science frameworks are not tools so much as *practices*, enacted by those who practice them, and they are necessarily changed in their uptake. As we took up CHAT in writing studies, we translated it for our own localized needs: we interpreted CHAT through case study methodology, synthesized it with genre theory, and kept it at a scale that allowed us to describe and examine interrelations among multiple textual mediators. In taking up the innovation of CHAT, we innovated further, changing it to meet our field's needs. In fact, if we had not been able to adapt CHAT for our needs, we would not have been persuaded to take it up in the first place. CHAT became persuasive to us because we could alter it to fit our network of associations. This translation was arguably a betrayal of what we had received—but no more so than the other translations enacted by our predecessors and contemporaries in other social science fields, in other associative networks, working on other projects.

In turn, we in writing studies might ourselves feel betrayed by later CHAT translations, translations that (as we saw in chapters 8–9) have resulted in larger scope. Whereas CHAT in the mid-1990s seemed custom-built for our concerns, its later translations no longer account for the more granular analyses of interrelated workplace texts that are important to our work; this level of description disappears in later Engeström-inflected CHAT, which focuses on creating agreements to govern ongoing relationships as opposed to describing intertextual mediation at a more granular level. I'm not saying our feelings are hurt—but that translation is not what we signed up for!

Which is fine. Engeström's CHAT is *a* social science approach for understanding work and workplaces in practice, *a* practice theory among others. Social science approaches such as CHAT are just as vulnerable to such inevitable translations and betrayals as the hard sciences are. As a "sociology of the social" (Latour 2006), the sociology of translation is well positioned to examine these betrayals, these rhetorical uptakes.

Throughout this book's account of CHAT's translations and betrayals, I asked, How has this approach changed as it has been taken up in different places and times for different purposes? What can the story of Engeström's CHAT tell us about how such social science frameworks develop over time?

And throughout the book, I have attempted to deploy not just analysis, not just critique, but also sympathy and understanding. Engeström's CHAT did not have to develop in this way. The translations that I have documented here have unfolded against a backdrop of variable, changing needs. In responding to these variations, CHAT has developed features that made sense at certain points in its history but sometimes create contemporary problems. Like a country's constitution, CHAT has been revised and reinterpreted in different eras, and that history has been written into the fabric of CHAT itself, in its layers of sedimentation, in its ongoing settlements.

As reviewed in chapter 9, Engeström's CHAT is currently attempting to form a new settlement, one that can allow it to address problems at the public scale. This latest settlement has not yet solidified, and it's unclear whether it will hold. I can't predict the future of Engeström's CHAT. But I can speculate, based on challenges that we are beginning to see in the literature.

So, in this final chapter, I'll review what we have learned about this approach through a translation analysis. In chapter 1, I asked these questions:

- How did Engeström's CHAT become Engeström's CHAT? That is, how did this variant of cultural-historical activity theory emerge through successively addressing different problems?
- What does this story tell us about how social science frameworks spread?
- What lines of critique, including self-critique, can this analysis open for us?

Let's take these in order.

HOW DID ENGESTRÖM'S CHAT EMERGE THROUGH SUCCESSIVELY ADDRESSING DIFFERENT PROBLEMS?

In chapter 3, we reviewed three metaphors for understanding development in CHAT, each of which emphasized a different kind of fidelity. The *generations* metaphor emphasizes fidelity to its lineage in a story of inevitable progress. The *strands* metaphor emphasizes fidelity and coherence within each strand, with the relatively ecumenical possibility that the strands are compatible enough to sometimes be woven together. And the *waves* metaphor emphasizes fidelity to source materials: each wave deposits additional facts that can be used to counteract betrayals, which are attributed to incompetence or deliberate misrepresentation.

Fidelity dreams of *diffusion*, in which an idea is so self-evidently good that it is picked up and transported without transformation (Akrich et al., 2002a). In being unchanged, it changes the world around it: in being taken up in different locales, it reconfigures how people in each locale understand the world. This dream of diffusion is the dream of inventors, scientists, engineers, and missionaries. It is the faith that—for instance—Vygotskian theory can change the world.

Diffusion is the dream. But we must live in the waking world, and that world changes the idea as much as the idea changes the world. Ideas do not travel by diffusion but by translation, taken up only if they can be understood to solve existing problems, and they are transformed in the service of those localized problems. To be taken up, an idea must be altered to address the concerns of those who are to take it up. This is true of inventors (Rogers, 2003), scientists (Latour, 1988), engineers (Latour, 1996), and missionaries (Comaroff & Comaroff, 1991) alike.

We saw this story play out repeatedly in part II. There, we explored the question of how Engeström's CHAT developed. We systematically explored how CHAT was taken up at specific points and translated: in terms of its empirical focus, its understanding of mastery, and the mediators it studied. Each deployment was applied to different problems: the New Human, the ideal (supervised) Soviet citizen, the emancipation of labor in fast-changing work environments, the reorganization of interorganizational work, and public policy implementation. In each translation, problematizations, interessements, enrollments, and mobilizations changed CHAT (table 12.1).

We also saw these points of translation playing out in miniature in chapter 7, where we examined the localized, heterodox translation of Engeström's CHAT for my own field of writing studies.

And as we also saw in part II, each point of translation also involved betrayal. Earlier CHAT components, concepts, and orientations were changed or dropped when they failed to support the current problem space. For instance, the central concept of the New Human that so animated Vygotsky's work was forgotten by the time CHAT came to the West—and it seems bizarre and unrealistic to contemporary practitioners of Engeström's CHAT, whose focus has shifted instead to workplace supermediators such as care agreements. (Such supermediators may have seemed similarly bizarre to Vygotsky and Luria in 1930.) These betrayals mean that CHAT might seem fundamentally different, a "mutant and hybrid" (Yasnitsky, 2016a, p. 225), compared to

Table 12.1

Problematization, interessement, enrollment, and mobilization in versions of CHAT

	Chapter 4: Vygotsky's CHAT	Chapter 5: Leontiev's CHAT	Chapter 6: Westernized CHAT in 1987	Chapter 8: Engeström's third-generation CHAT	Chapter 9: Engeström's fourth-generation CHAT
Problematization	Pre-Marxist psychology	Psychology that did not fit the strictures of Marxism, *partiinost'*, and practicality	Information-processing cognitive psychology, design	Managerial and social theory	Public policy initiatives
Interessement	Psychologists, pedologists, teachers, special educators, and others	Psychologists, rehabilitated soldiers, children, university students, and the Communist Party	Unions, management, academic researchers, and the broader Scandinavian commitment to workplace democracy	Sociologists, management scholars, vocational education researchers, institutional funders, and institutional participants	Public officials and agencies, public workers, the homeless, communities, institutional funders, CHAT scholars
Enrollment	Interested psychologists via academic publications, popular books, pedagogical materials, a pedology network, and both academic and popular lectures	Rubinshtein, Pavlov, Lysenko, others in the Marxist science apparatus	CHAT forebears (Vygotsky, Leontiev, Davydov); Western cultural psychologists; Jungk and other Europeans focused on transformational participation	Barley, Deleuze & Guattari, Giddens, Victor & Boynton, other managerial and social theorists	Finnish Housing First
Mobilization	Psychology, pedology	Psychology, physiology	Interventionist design research across fields	Organizational transformation	Policy compliance

Table 12.2
Concepts attributed to Vygotsky or Leontiev and how they might have been attributed if Luria and Leontiev had not been such energetic promoters

Accomplishments attributed to Vygotsky or Leontiev	Equally truthful attributions
The principle of double stimulation	Rooted in Köhler, Basov, Ach, and Janet; invented by Sakharov
The zone of proximal development	Conceived by Bergsen; developed by McCarthy, inspired by Lewin's "field theory"
Mediation	Based in Pavlov's "second signal system"
The unit of analysis	Extrapolated by Basov from Köhler's writings
The concept of activity	Borrowed by Rubinshtein from philosophy, revised by him to work in psychology, and legitimized through his reading of Engels

Vygotsky's instrumentalist work. But without those translations, CHAT might not have spread so successfully. Vygotsky, were he alive today, would experience both the pride of providing a foundation for this later work and perhaps the horror of seeing what has been built on it.

Indeed, even as Vygotsky is invoked and cited in contemporary work, the concepts and arguments attributed to him—including double stimulation, the zone of proximal development, and even units of analysis—have been translated from his early Marxist psychology to support Engeström's CHAT as organizational sociology and as public policy. The terms *perform* fidelity to the past, but without translation, they would have stayed in the past. Without such translations, starting with Vygotsky's systematic and partial rehabilitation by Luria and Leontiev in the late 1950s, it's possible that Vygotsky himself would be largely lost to the past, just as Basov, Shpet, and Potebnya have been. If Luria and Leontiev had not promoted Vygotsky in the late 1950s, if Luria had not twisted Michael Cole's arm about publishing an English-language collection of Vygotsky's writings, if Leontiev had been less nimble about promoting activity theory, Vygotsky might easily have been a minor figure in psychology. The accomplishments attributed to him could have been easily—and truthfully—attributed to someone else (table 12.2).

The point here is not that Vygotsky and Leontiev were empty suits. They did crucial and important work in problematizing questions, interesting others in these questions, and enrolling and mobilizing allies—but *in addition*, others found it useful to portray them as obligatory passage points. When Leontiev and Luria attributed accomplishments to Vygotsky, they told a story that elevated their own work. Still others attributed accomplishments to Leontiev (rather than Rubinshtein, for instance), simplifying the story they told in their turn. And when Engeström and Sannino applied double stimulation and the zone of proximal development to collectives rather than individuals, they portrayed their work as continuing Vygotsky's accomplishments. Great and triumphant figures in history are constructed through such attributions because simple stories of triumph are easy to tell and understand. This is why Engeström's story of CHAT's generations has been so powerful in spreading Engeström's CHAT.

WHAT DOES THIS STORY TELL US ABOUT HOW SOCIAL SCIENCE FRAMEWORKS SPREAD?

Examining these translations side by side in part III, we see the impact of theoretical and methodological decisions on Engeström's CHAT. In particular, the activity system, which seems like such a fundamental assumption in the analysis, can be understood as an analytical construct that organizes the data a researcher collects and analyzes. Without this unit of analysis, researchers can and do investigate cases differently, generate different findings, and conclude different things. Such insights are not new to those of us who have worked with qualitative research methodology, but they are thrown into sharp relief as we consider how CHAT researchers have applied the concept and term of "unit of analysis."

More broadly, this story demonstrates how we create and mobilize constructs to help us conceptualize, bound, and analyze social phenomena such as workplaces. As we saw in chapter 11, constructs such as the unit of analysis—constructs that are represented by the many triangles we saw in chapter 1—are based on assumptions about what is important in a social science investigation: which people, which things, which relations. Like maps, they simplify the territory, showing only the features that a presumed reader needs for a given sort of navigation. Such maps are never totalizing. The territory is always much more complicated than the map. And that is what makes maps work so well: by simplifying the territory, a map becomes combinable and superimposable with other maps, yielding new views and insights.

Yet, as we saw in chapter 10, Engeström's CHAT has consistently sought to make its maps more complicated. It has chased the ever-receding horizon of "context," seeking a unit of analysis that is the minimal unit of all social relations within a case. Since the map is always less detailed than the territory it represents, Engeström's CHAT has proffered a series of units of analysis to progressively expand its analysis: from an activity system to linked activity systems to activity networks to coalescing cycles of expansive learning. These ever-larger units of analysis correspond to ever-larger empirical scope, and arguably, its empirical and analytical methods have not caught up.

Social phenomena thus overflow the available units of analysis and units of observation we have at our disposal. No one has produced, or can produce, a totalizing construct that can equate the map with the territory. Yes, Leontiev argued that there was one best methodological path (Levitin, 1982, p. 112), but that path relied on the silent partner of the state and the relentless universal logic of Stalinist dialectics. Today, we can rely on neither one of these, and without a universal extra-scientific arbiter such as the Soviet state, meaning must emerge from dialogue (Spinuzzi, 2023a).

This brings us to the question of arbiters. In *Developmental Psychology in the Soviet Union* (1988), Valsiner applies Kuhn's work on scientific paradigms to explore how developmental psychologists in the USSR advanced their field by political as well as scientific means. He emphasizes that in the USSR, the Western split between science and society was not established: Soviet philosophers and psychologists freely used "extra-scientific authorities as arbiters in their relationships" (p. 11; cf. Bauer, 1952, p. 4). This is true and demonstrable. But Latour might counter that the USSR was simply more *open* about using such arbiters. Western studies in science and technology have exposed political and cultural influences as well. Consider the fact that Pasteur had to make public demonstrations (Latour, 1988) and that Edison had to cultivate a public persona of genius (Bazerman, 1999), among many other examples. Consider also the political priorities that govern research funding. Science and society are not split: social science is conducted in and through social interests, and its value is interpreted through social effects. If social science approaches spread through translation, we must understand the extra-scientific interests that influence the rhetorical uptake—the translation and the betrayal—of these approaches.

WHAT LINES OF CRITIQUE, INCLUDING SELF-CRITIQUE, CAN THIS ANALYSIS OPEN FOR US?

These insights lead us to the third question asked in chapter 1: What lines of critique, including self-critique, can this analysis open for us? As I have argued elsewhere (Spinuzzi 2008), Engeström's CHAT does not support a strong political-rhetorical critique. It has instead developed an apolitical, managerial stance, originating in Leontiev's savvy bureaucratic maneuvering (see chapter 5). This stance was perhaps critical to CHAT's survival, but it limits the ability of Engeström's CHAT to level political or rhetorical critiques. It makes invisible the influence of extra-scientific authorities and public interests, or it domesticates these influences as outputs that Engeström's CHAT can manage (see chapter 9). Specifically, Engeström's CHAT has an institutional bias tied to its focus on organizations as settings and its interventionism, which is grounded in internal critique within a given institution. This bias becomes most obvious in the public policy work in chapter 9, in which workers are expected to accept and implement changes in accordance with the public policy, but not question it.

The interactions in change laboratories are meant to be a deliberative *dialectic*, a conversation that leads to unity, instantiated in a single monologic agreement. Examples include care agreements, mobility agreements, rental agreements, negotiated care plan documents, and other coordinative genres on which everyone agrees and to which everyone is held (Engeström, 2023). But as CHAT has been applied to interindividual, interorganizational, and public cases, fewer mechanisms exist for enforcing or incentivizing such agreements (Spinuzzi 2020). Instead, what is needed is *dialogue* in which meaning emerges from difference rather than unity.

Yet Matusov argues that a tension exists between dialogue and activity: "activity is responsible for the monologicity aspect of discourse" because "joint collective activity is about accomplishing something" and

> the subject of such an activity is a unified, shared, common understanding—one consciousness, as Bakhtin would say. A joint activity becomes problematic when shared understanding is not achieved, partially achieved, or achieved about wrong things. Although heteroglossia can be viewed as a productive force in the activity at its initial and intermediary stages, at the final phase, it has to be eliminated. From this point of view, *activity is essentially anti-dialogue* (anti-heteroglossic). However, as Bakhtin showed, this unifying, centripetal force is an important aspect of any discourse defining one's voice, the recognized unity of consciousness. The problem starts when the other complementary and necessary aspect of discourse—namely dialogicity—is either ignored or attempted to actively exclude from the analysis (and design) or eliminate from the discourse, when a voice becomes the voice. In the latter case, *there becomes a tendency to establish a regime of excessive monologism.* (Matusov, 2013, p. 383, my emphasis)

Matusov praises the fact that Engeström's CHAT has incorporated dialogue. But that dialogue tends to be ultimately in service of dialectic. When the endpoint is an agreement that an institutional authority can enforce, it becomes difficult to develop an internal critique. But as work becomes more distributed, more projectified (Spinuzzi 2015a), such institutional authorities are in short supply: few extra-scientific authorities exist who can create a forced perspective or serve as a final arbiter. Without institutional pressure, it becomes difficult to even delineate a workplace (Spinuzzi, 2023b). The closest we can get to such authorities is settled public policy, which is perhaps why Engeström's CHAT took that turn in chapter 9.

With these points in mind, let's return to the metaphors of generations, strands, and waves in chapter 3. These metaphors each have a relationship with fidelity: changes are interpreted in terms of whether they perform fidelity to a lineage, a strand, or a source. But fidelity carried too far is *preservation*, like the preservation of an insect in amber. If we want to preserve a theory unchanged, protected from decay, we must make it inert—dead. But if we want it to live, we must understand changes in terms of dialogue, not dialectic: in how meaning emerges from differences, not unity. We must see translations, not as betrayals that kill a theory, but as ways for the theory to become meaningful to others.

We began this book by looking at the seemingly vast, inexplicable gulf between two triangles, drawn fifty-seven years apart, and wondering what they could tell us about the ongoing family fight in CHAT. But the good news is that we don't need to fight: there is room enough for plenty of triangles (and other shapes) in this interdisciplinary social science approach called CHAT—if only we decide to make room for them.

NOTES

CHAPTER 1

1. Throughout the text, I use the most common English spellings for Russian names, such as Vygotsky and Leontiev. Some sources use different spellings, such as Vygotskii and Leontyev, so I use those spellings in the citations.

2. Here and throughout, I use "phenomenon" in the generic sense (as an observable or detectable fact) rather than in the sense used in phenomenology. In this generic sense, phenomena are made visible through inscriptions (Latour 1986). Through these inscriptions, a phenomenon is stabilized enough for actors to interdefine each other. "The human sciences do not show up as the curtain falls, in order to interpret the phenomenon. They constitute the phenomenon" (Latour, 1996, p. 180). This understanding is explicitly opposed to that of phenomenology (Latour, 2006, p. 61).

CHAPTER 2

1. This question of methodology has been discussed frequently. Veresov (2010) argues that Vygotsky's small-scale studies reflected his move from a descriptive to an explanatory paradigm. Mahn (2010) adds that Vygotsky preferred experiments to be more ecological, less divorced from the real world. Valsiner (1988, pp. 2–3) concedes that American readers look for data in Vygotsky and cannot find any—because instead of measuring the outcome, Vygotsky was more interested in describing the process leading to such encounters (p. 137). According to Valsiner, this process description substituted for the dependent variable in traditional psychological experiments. In this sense, Vygotsky follows Köhler (1917/2013), whom he frequently cites. Yet in taking this tack, Vygotsky and collaborators sometimes read their assumptions into the research itself, as is most obvious in their Uzbek studies.

CHAPTER 4

1. English translations of Vygotsky's and others' Russian works often use "man" or "Man" to refer to humanity in general, but the Russian term denotes humanity rather than a specific gender. In this book, I substitute nongendered terms such as "human" when referring to human capabilities. Thanks to Bonnie Nardi and Victor Kaptelinin for suggesting this substitution.

2. Tomasello (2019) argues that "internalization is nothing other than role-reversal imitation used in a flexible way: the child imitates others directing her behavior, or, alternatively, imitates herself teaching others, with herself substituted as learner" (p. 153).

3. This passage was popular among Soviet psychologists: it was also paraphrased by Rubinshtein in 1935 (Payne 1968, p. 90) and Leontiev in 1959 (Leontyev 1959/1981).

4. Karpov & Haywood (1998) argue that mediation can be further separated into metacognitive mediation ("the acquisition of semiotic tools of self-regulation") and cognitive mediation ("the acquisition of scientific concepts representing the essence of some class of phenomena") (p. 27).

CHAPTER 5

1. Although Bakhurst (2009) and Levant (2018) later criticized Engeström for turning activity theory toward the practical, Engeström did not initiate that practical turn: it happened here, in the work of Leontiev and his Kharkov Circle, partly in response to the Great Terror and the strictures of Stalinist science.

2. Thanks to Dmitry Leontiev and Anton Yasnitsky for independently confirming the original date of this publication.

3. Luria used Pavlov's language as a cover to carry on Vygotsky's third-stage holistic research. Pavlov argued shortly before he died that "language acts as a 'second signal system'" (Cole & Maltzman 1969, p. 10); this endorsement was a boon to psychologists working on "verbal behavior and language learning" (p. 10), including Luria, who repositioned Vygotsky and his collaborators as working with Pavlov's second signal system (Luria 1957b, 1966). As Derry explains, "The second signal system incorporates the notion that language and concepts mediate human existence as a second signal system rather than as a first signal where stimuli act on the nervous system directly" (2013, p. 98). Early in his career, Vygotsky built on Pavlov, characterizing consciousness as a reflex of reflexes (1997c); late in his career, Pavlov used a similar formation, characterizing words as signals of signals (Payne 1968). This formulation allowed Luria to rehabilitate Vygotsky's work under the cover of following Pavlov.

CHAPTER 6

1. The levels of activity had previously been mentioned by Western cultural psychologists working in the Vygotskian tradition, such as Lave et al. (1984) and Wertsch (1985), but Bødker made them central to her empirical analysis.

CHAPTER 7

1. This history draws on Spinuzzi (2021b).

CHAPTER 8

1. Thanks to an anonymous reviewer for suggesting that I examine funding sources.

CHAPTER 9

1. This chapter is based in part on (Spinuzzi, 2021a).

2. Sannino characterizes these studies as activist and focused on the betterment of Vygotsky's and Luria's participants. Yet Luria's work with the dynamoscope involved finding people at points of intense vulnerability (waiting to find out if they will be purged from the university or arrested but without knowing the charges) and forcing them to undergo experiments that emphasized that stress. One study additionally involved implanting false memories of shameful deeds into the minds of innocent participants via hypnosis in order to study how they would handle this tension (possibly discovering a sociopath along the way: 1960, p. 144). His twin study was not set up to help both twins, as Sannino argues (2011, p. 588), but rather, to provide intense support for one and leave the other *untreated* as a control; the second twin's (lesser) improvement was an unintended side effect (see chapter 4, this book). Similarly, his cross-cultural work involved drafting Uzbeks into his study under the implied threat of security services (Lamdan & Yasnitsky 2016; see chapter 11, this book). Additionally, Luria, Leontiev, and Vygotsky all conducted research under the aegis of their research institutes, and it's not clear that their efforts were motivated by activism rather than their institutional positions. Vygotsky's writings focused on elevating extraordinary abilities to yield the New Human (Vygotsky 1930/1994b; Vygotsky & Luria 1930/1993, 1930/1994; Luria 1968; cf. Yasnitsky 2018b), even as his official duties focused on children with developmental challenges and disabilities. I am not criticizing Vygotsky, Luria, and Leontiev for working with a different set of professional and cultural expectations than we currently have in Western social science—rather, I am arguing that *we* should be cautious about projecting our current ideals and ideas about activism back into their work. For a prolonged argument about Vygotsky's commitment to an activist stance, see Stetsenko (2017).

3. An anonymous reviewer with design expertise remarks that this process does not seem well developed from a design perspective. Specifically, wicked problems resist simple solutions, requiring participants to engage in open-ended dialogue. Yet the second stimuli are simple solutions that will not help the participants to lend their expertise or fundamentally transform the problem. I agree. Although these sessions *imitate* design sessions, they address lower-level epistemic questions, encouraging participants to *accept* policy decisions rather than actually deliberate over or design them.

CHAPTER 11

1. Lamdan & Yasnitsky (2016) cover this story well and in far more detail, and their chapter provided me with the citations to Harrower and to Koffka's addition to Luria's 1934 report. However, the analysis in this chapter is mine, based on the source materials.

2. As noted earlier, in Luria's early work, he had few compulsions about compelling people to participate in his studies (see Luria 1960).

WORKS CITED

Acquier, A., Daudigeos, T., & Pinkse, J. (2017). Promises and paradoxes of the sharing economy: An organizing framework. *Technological Forecasting and Social Change, 125*(July), 1–10. https://doi.org/10.1016/j.techfore.2017.07.006

Adler, P. S., & Heckscher, C. (2007). Towards collaborative community. In C. Heckscher & P. S. Adler (Eds.), *The firm as a collaborative community: Reconstructing trust in the knowledge economy* (pp. 11–105). Oxford University Press.

Akrich, M. (1993). A gazogene in Costa Rica: An experiment in techno-sociology. In P. Lemonnier (Ed.), *Technological choices: Transformation in material cultures since the Neolithic* (pp. 289–337). Routledge.

Akrich, M., Callon, M., & Latour, B. (2002a). The key to success in innovation part I: The art of interessement. *International Journal of Innovation Management, 6*(2), 187–206.

Akrich, M., Callon, M., & Latour, B. (2002b). The key to success in innovation part II: the art of choosing good spokespersons. *International Journal of Innovation Management, 6*(2), 207–255.

Andersen, J., Bazerman, C., & Schneider, J. (2014). Beyond single genres: Pattern mapping in global communication. In E.-M. Jakobs & J. Perrin (Eds.), *Handbook of writing and text production* (pp. 305–322). De Gruyter.

Arievitch, I. M. (2017). *Beyond the brain: An agentive activity perspective on mind, development, and learning.* Sense Publishers.

Artemeva, N. (2005). A time to speak, a time to act: A rhetorical genre analysis of a novice engineer's calculated risk taking. *Journal of Business and Technical Communication, 19*(4), 389–421.

Artemeva, N. (2008). Toward a unified social theory of genre learning. *Journal of Business and Technical Communication, 22*(2), 160–185.

Artemeva, N. (2009). Stories of becoming: A study of novice engineers learning genres of their profession. In C. Bazerman, A. Bonini, & D. Figueiredo (Eds.), *Genre in a changing world* (pp. 158–178). WAC Clearinghouse.

Artemeva, N. (2019). An engrained part of my career: The formation of a knowledge worker in the dual space of engineering knowledge and Rhetorical process. In C. Bazerman, B. Y. G. Pinzón, D. Russell, P. Rogers, L. B. Peña, E. Narváez, P. Carlino, & M. Tapia (Eds.), *Conocer la escritura: Investigación más allá de las fronteras [Knowing writing: Writing research across borders* (pp. 321–350). Pontificia Universidad Javeriana.

Artemeva, N., & Freedman, A. (2001). "Just the boys playing on computers": An activity theory analysis of differences in the cultures of two engineering firms. *Journal of Business and Technical Communication, 15*(2), 164–194.

Artemeva, N., & Freedman, A. (Eds.). (2007). *Rhetorical genre studies and beyond* (p. 282). Inkshed Publications.

Arunkumar, M., Van Paridon, J., Ostarek, M., & Huettig, F. (2021). Do illiterates have illusions? A conceptual (non) replication of Luria (1976). *Journal of Cultural Cognitive Science, 5*(2), 143–158.

Arvola, M. (2004). The interaction character of computers in co-located collaboration. In E. O'Neill, P. Palanque, & P. Johnson (Eds.), *People and computers XVII—designing for society* (pp. 37–51). Springer.

Arvola, M. (2014). Interaction and service design as offering perspectives in a space of action. In Y. Lim, K. Niedderer, J. Redström, E. Stolterman, & A. Valtonen (Eds.), *Design's big debates: Proceedings of the 2014 Design Research Society (DRS) conference* (pp. 7–15). Design Research Society.

Avis, J. (2009). Transformation or transformism: Engestrom's version of activity theory? *Educational Review, 61*(2), 151–165. https://doi.org/10.1080/00131910902844754

Bakhtin, M. M. (1981). *The dialogic imagination: Four essays*. University of Texas Press.

Bakhurst, D. (2009). Reflections on activity theory. *Educational Review, 61*(2), 197–210. https://doi.org/10.1080/00131910902846916

Bal, A., Bear, A. B., Ko, D., & Orie, L. (2021). Indigenous learning lab: Inclusive knowledge-production and systemic design toward indigenous prolepsis. In W. Cavendish & J. F. Samson (Eds.), *Intersectionality in education: Toward more equitable policy, research, and practice* (pp. 1317–1320). Teachers College Press.

Bannon, L. J., & Bodker, S. (1991). Beyond the interface: Encountering artifacts in use. In J. M. Carroll (Ed.), *Designing interaction: Psychology at the human-computer interface* (pp. 227–253). Cambridge University Press.

Bansler, J. P. (1989). Systems development research in Scandinavia: Three theoretical schools. *Scandinavian Journal of Information Systems, 1*(1), 3–20.

Barley, S. R. (1988). On technology, time and social order: Technically induced change in the temporal organization of radiological work. In F. A. Dubinskas (Ed.), *Making time: Ethnographies of high-technology organizations* (pp. 123–169). Temple University Press.

Basit, T. N. (2003). Manual or electronic? The role of coding in qualitative data analysis. *Educational Research, 45*(2), 143–154.

Bateson, G. (1958). *Naven* (2nd ed.). Stanford University Press.

Bauer, R. (1952). *The new man in Soviet psychology*. Harvard University Press.

Bazerman, C. (1988). *Shaping written knowledge: The genre and activity of the experimental article in science*. University of Wisconsin Press.

Bazerman, C. (1994). Systems of genre and the enactment of social intentions. In A. Freedman & P. Medway (Eds.), *Genre and the new rhetoric* (pp. 79–99). Taylor & Francis.

Bazerman, C. (1999). *The languages of Edison's light*. MIT Press.

Bazerman, C. (2004a). Intertextualities: Voloshinov, Bakhtin, literary theory, and literacy studies. In A. F. Ball & S. W. Freedman (Eds.), *Bakhtinian perspectives on language, literacy, and learning* (pp. 53–65). Cambridge.

Bazerman, C. (2004b). Speech acts, genres, and activity systems: How texts organize activity and people. In C. Bazerman & P. Prior (Eds.), *What writing does and how it does it: An introduction to analyzing texts and textual practices* (pp. 309–340). Lawrence Erlbaum Associates.

Bazerman, C. (2022). Revisiting the early uses of writing in society building: Cuneiform culture and the Chinese imperium. *Literatura y Linguistica, 46*, 61–76.

Bazerman, C. (2023). *How I became the kind of writer I became: An experiment in autoethnography*. WAC Clearinghouse.

Bazerman, C., Little, J., & Chavkin, T. (2003). The production of information for genred activity spaces: Informational motives and consequences of the environmental impact statement. *Written Communication, 20*(4), 455–477. https://doi.org/10.1177/0741088303260375

Bazerman, C., & Russell, D. R. (Eds.). (2003). *Writing selves/writing societies: Research from activity perspectives*. WAC Clearinghouse.

Becker, H. S. (2008). *Tricks of the trade: How to think about your research while you're doing it*. University of Chicago Press.

Beniger, J. (1989). *The control revolution: Technological and economic origins of the information society*. Harvard University Press.

Benkler, Y. (2006). *The wealth of networks*. Yale University Press.

Berkenkotter, C. (1991). Paradigm debates, turf wars, and the conduct of sociocognitive inquiry in composition. *College Composition and Communication, 42*(2), 151–169.

Berkenkotter, C., & Huckin, T. N. (1995). *Genre knowledge in disciplinary communication: Cognition/culture/power*. Lawrence Erlbaum Associates.

Berlin, J. (1988). Rhetoric and ideology in the writing class. *College English, 50*(5), 477–494.

Bernard, H. R. (2002). *Research methods in anthropology: Qualitative and quantitative approaches* (3rd ed.). AltaMira Press.

Bernard, H. R. (2011). *Research methods in anthropology: Qualitative and quantitative approaches* (5th ed.). Rowman & Littlefield.

Bertelsen, O. W., & Bødker, S. (2000). Information technology in human activity. *Scandinavian Journal of Information Systems, 12*(1), 3–14.

Bertelsen, O. W., & Bødker, S. (2003). Activity theory. In J. M. Carroll (Ed.), *HCI models, theories, and frameworks: Toward a multidisciplinary science* (pp. 291–324). Morgan Kauffman.

Beyer, H., & Holtzblatt, K. (1998). *Contextual design: Defining customer-centered systems*. Morgan Kaufmann.

Bjerknes, G. (1992). *Shared responsibility: A field of tension*. Springer-Verlag.

Bjerknes, G., & Bratteteig, T. (1995). User participation and democracy: A discussion of Scandinavian research on system development. *Scandinavian Journal of Information Systems*, *7*(1), 73–98.

Bjerknes, G., Ehn, P., & Kyng, M. (Eds.). (1987). *Computers and democracy: A Scandinavian challenge*. Avebury.

Blakeslee, A. M. (1997). Activity, context, interaction, and authority: Learning to write scientific papers in situ. *Journal of Business and Technical Communication*, *11*(2), 125–169.

Blunden, A. (2010). *An interdisciplinary theory of activity*. Brill.

Blunden, A. (Ed.). (2014a). *Collaborative projects: An interdisciplinary study*. Brill.

Blunden, A. (2014b). Conclusion: Time and activity. In A. Blunden (Ed.), *Collaborative projects: An interdisciplinary study* (pp. 368–374). Brill.

Blunden, A. (2014c). Introduction: "Collaborative project" as a concept for interdisciplinary human science research. In A. Blunden (Ed.), *Collaborative projects: An interdisciplinary study* (pp. 1–28). Brill.

Blunden, A. (2020). The unit of analysis and germ cell in Hegel, Marx and Vygotsky. Ethical Politics. https://www.ethicalpolitics.org/ablunden/pdfs/unit_and_cell.pdf

Blunden, A., & Arnold, M. (2014). Formation of the concept of "collaborative learning space." In A. Blunden (Ed.), *Collaborative projects: An interdisciplinary study* (pp. 158–175). Brill.

Blyler, N. R., & Thralls, C. (Eds.). (1993). *Professional communication: The social perspective* (p. 292). Sage.

Bødker, S. (1987). *Through the interface: A human activity approach to user interface design*. Ph.D. thesis, Aarhus Universitet. http://pure.au.dk/portal/files/20861897/Full_text

Bødker, S. (1989). A human activity approach to user interfaces. *Human-Computer Interaction*, *4*(3), 171–195.

Bødker, S. (1991). *Through the interface: A human activity approach to user interface design*. Lawrence Erlbaum Associates.

Bødker, S. (1997). Computers in mediated human activity. *Mind, Culture, and Activity*, *4*(3), 149–158.

Bødker, S. (2009). Past experiences and recent challenges in participatory design research. In A. Sannino, H. Daniels, & K. D. Gutierrez (Eds.), *Learning and expanding with activity theory* (pp. 274–285). Cambridge University Press.

Bødker, S., & Andersen, P. B. (2005). Complex mediation. *Human-Computer Interaction, 20*(4), 353–402.

Bødker, S., Ehn, P., Kammersgaard, J., Kyng, M., & Sundblad, Y. (1987). A UTOPIAN experience: On design of powerful computer-based tools for skilled graphical workers. In G. Bjerknes, P. Ehn, & M. Kyng (Eds.), *Computers and democracy: A Scandinavian challenge* (pp. 251–278). Avebury.

Bødker, S., Greenbaum, J., & Kyng, M. (1991). Setting the stage for design as action. In J. Greenbaum & M. Kyng (Eds.), *Design at work: Cooperative design of computer systems* (pp. 139–154). Lawrence Erlbaum Associates.

Bødker, S., & Iversen, O. S. (2002). Staging a professional participatory design practice: Moving PD beyond the initial fascination of user involvement. *Proceedings of the second Nordic conference on human-computer interaction, 31*, 11–18. https://doi.org/10.1145/572020.572023

Bødker, S., & Klokmose, C. N. (2011). The human-artifact model: An activity theoretical approach to artifact ecologies. *Human–Computer interaction, 26*(4), 315–371. https://doi.org/10.1080/07370024.2011.626709

Bødker, S., & Klokmose, C. N. (2012). Dynamics in artifact ecologies. *Proceedings of the 7th Nordic conference on Human-computer interaction: Making sense through design*, 448–457. https://doi.org/10.1145/2399016.2399085

Bødker, S., & Petersen, M. G. (2000). Design for learning in use. *Scandinavian Journal of Information Systems, 12*(1), 61–80.

Boisot, M., Nordberg, M., Yami, S., & Nicquevert, B. (Eds.). (2011). *Collisions and collaboration: The organization of learning in the ATLAS experiment at the LHC*. Oxford University Press.

Boucke, O. F. (1923). The limits of social science. II. *American Journal of Sociology, 28*(4), 443–460.

Bracewell, R. J., & Witte, S. P. (2003). Tasks, ensembles, and activity: Linkages between text production and situation of use in the workplace. *Written Communication, 20*(4), 511–559.

Brown, J. S., Heath, C., & Pea, R. (1999). *Perspectives on activity theory*. Cambridge University Press.

Brozek, J. M. (1972). Noted figures in the history of Soviet psychology: Pictures and brief biographies. In J. M. Brozek (Ed.), *Psychology in the USSR: An historical perspective* (pp. 22–29). International Arts and Sciences Press.

Bruni, A., Gherardi, S., & Parolin, L. L. (2007). Knowing in a system of fragmented knowledge. *Mind, Culture, and Activity, 14*(1–2), 83–102.

Byrkjeflot, H. (2001). *The Nordic model of democracy and management*. Fagbokforlaget.

Callon, M. (1986a). Some elements of a sociology of translation: Domestication of the scallops and the fishermen of Saint Brieuc Bay. In J. Law (Ed.), *Power, action and belief: A new sociology of knowledge?* (pp. 67–83). Routledge.

Callon, M. (1986b). The sociology of an actor-network: The case of the electric vehicle. In M. Callon, J. Law, & A. Rip (Eds.), *Mapping the dynamics of science and technology: Sociology of science in the real world* (pp. 19–34). Macmillan.

Callon, M., Lascoumes, P., & Barthe, Y. (2009). *Acting in an uncertain world: An essay on technical democracy*. MIT Press.

Carroll, J. M. (1990). *The Nurnberg funnel: Designing minimalist instruction for practical computer skill*. MIT Press.

Carroll, J. M. (1991). Introduction: The Kittle House manifesto. In John M. Carroll (Ed.), *Designing interaction: Psychology at the human-computer interface* (pp. 1–16). Cambridge University Press.

Castells, M. (2003). *The Internet galaxy: Reflections on the Internet, business, and society*. Oxford University Press.

Charney, D. (1998). From logocentrism to ethocentrism: Historicizing critiques of writing research. *Technical Communication Quarterly*, *7*(1), 9–32. https://doi.org/10.1080/10572259809364615

Christiansen, E. (1996). Tamed by a rose: Computers as tools in human activity. In B. A. Nardi (Ed.), *Context and consciousness* (pp. 175–198). MIT Press.

Cohen, L., Manion, L., & Morrison, K. (2000). *Research methods in education* (5th ed.). Routledge.

Cole, M. (1996). *Cultural psychology: A once and future discipline*. Belknap Press of Harvard University Press.

Cole, M., & Engeström, Y. (1993). A cultural-historical approach to distributed cognition. In G. Salomon (Ed.), *Distributed cognitions: Psychological and educational considerations* (pp. 1–46). Cambridge University Press.

Cole, M., & Maltzman, I. (Eds.). (1969). *A handbook of contemporary Soviet psychology*. Basic Books.

Cole, M., & Scribner, S. (1974). *Culture and thought: A psychological introduction*. Wiley.

Collins, C. S., & Stockton, C. M. (2018). The central role of theory in qualitative research. *International Journal of Qualitative Methods*, *17*(1), 1–10.

Colvin, J., & Mukute, M. (2018). *Governance in Ethiopia: Impact evaluation of the African climate change and resilience Alliance (ACCRA) project*. Oxfam. https://doi.org/10.21201/2017.1756

Comaroff, J., & Comaroff, J. L. (1991). *Of revelation and revolution: Vol. 1. Christianity, colonialism, and consciousness in South Africa*. University of Chicago Press.

Coney, M. B., & Chatfield, C. S. (1996). Rethinking the author-reader relationship in computer documentation. *Journal of Computer Documentation*, *20*(2), 23–29.

Corbin, J. M., & Strauss, A. (1990). Grounded theory research: Procedures, canons, and evaluative criteria. *Qualitative Sociology*, *13*(1), 3–21.

Creswell, J. W. (2006). *Qualitative inquiry and research design: Choosing among five traditions*. SAGE.

Culhane, D. P., Metraux, S., Byrne, T., Stino, M., & Bainbridge, J. (2013). The age structure of contemporary homelessness: Evidence and implications for public policy. *Analyses of Social Issues and Public Policy, 13*(1), 228–244.

Daniels, H., Edwards, A., Engeström, Y., & Ludvigsen, S. R. (Eds.). (2010). *Activity theory in practice: Promoting learning across boundaries and agencies*. Routledge.

Daniels, H., Leadbetter, J., Warmington, P., Edwards, A., Martin, D., Popova, A., Apostolov, A., Middleton, D., & Brown, S. (2007). Learning in and for multi-agency working. *Oxford Review of Education, 33*(4), 521–538. https://doi.org/10.1080/03054980701450811

Decortis, F., Noirfalise, S., & Saudelli, B. (2000). Activity theory, cognitive ergonomics and distributed cognition: Three views of a transport company. *International Journal of Human-Computer Studies, 53*(1), 5–33. https://doi.org/10.1006/ijhc.2000.0378

de Laet, M., & Mol, A. (2000). The Zimbabwe bush pump: Mechanics of a fluid technology. *Social Studies of Science, 30*(2), 225–263.

Deleuze, G., & Guattari, F. (1987). *A thousand plateaus: Capitalism and schizophrenia*. University of Minnesota Press.

Derry, J. (2013). *Vygotsky: Philosophy and education*. Wiley Blackwell.

Dias, P., Freedman, A., Medway, P., & Pare, A. (1999). *Worlds apart: Acting and writing in academic and workplace contexts*. Lawrence Erlbaum Associates.

Ding, H. (2008). The use of cognitive and social apprenticeship to teach a disciplinary genre: Initiation of graduate students into NIH grant writing. *Written Communication, 25*(1), 3–52. https://doi.org/10.1177/0741088307308660

Di Russo, S. D. (2016). *Understanding the behaviour of design thinking in complex environments* [Dissertation, Swinburne University of Technology]. https://researchbank.swinburne.edu.au/file/a312fc81-17d3-44b5-9cc7-7ceb48c7f277/1/Stefanie%20Di%20Russo%20Thesis.pdf

Doody, S., & Artemeva, N. (2021). "Everything is in the Lab book": Multimodal writing, activity, and genre analysis of symbolic mediation in medical physics. *Written Communication, 39*(1), 3–43. https://doi.org/10.1177/07410883211051634

Dreyfus, H., & Dreyfus, S. E. (1986). *Mind over machine: The power of human intuition and expertise in the age of the computer*. Free Press.

Drucker, P. F. (1994). *Post-capitalist society*. Harper Paperbacks.

Dumez, H. (2015). What is a case, and what is a case study? *Bulletin of Sociological Methodology/Bulletin de Méthodologie Sociologique, 127*(1), 43–57.

Edwards, A. (2009). From the systemic to the relational: Relational theory and activity theory. In A. Sannino, H. Daniels, & K. D. Gutierrez (Eds.), *Learning and expanding with activity theory* (pp. 197–211). Cambridge.

Ehn, P. (1989). *Work-oriented design of computer artifacts*. Lawrence Erlbaum Associates.

Ehn, P., & Kyng, M. (1991). Cardboard computers: Mocking-it-up or hands-on the future. In *Design at work: Cooperative design of computer systems*, 169–196. Lawrence Erlbaum Associates.

Ekeblad, E. (1999). The emergence and decay of multilogue: Self regulation of a scholarly mailing list. *European Association for Research on Learning and Instruction (EARLI)*. https://lchc.ucsd.edu/mca/Paper/eva/The%20Dynamics%20of%20Multilogue.htm

Elhammoumi, M. (2001). Lost—Or merely domesticated? The boom in socio-historical theory emphasises some concepts, overlooks others. In S. Chaiklin (Ed.), *The theory and practice of cultural-historical psychology* (pp. 200–217). Aarhus University Press.

Engels, F. (1971). *Dialectics of nature*. International Publishers.

Engeström, Y. (1990). *Learning, working, and imagining: Twelve studies in activity theory*. Orienta-Konsultit Oy.

Engeström, Y. (1992). *Interactive expertise: Studies in distributed working intelligence*. University of Helsinki.

Engeström, Y. (1994). *Training for change: New approach to instruction and learning in working life*. International Labour Office.

Engeström, R. (1995). Voice as communicative action. *Mind, Culture, and Activity, 2*(3), 192–214.

Engeström, Y. (1996a). Developmental studies of work as a testbench of activity theory: The case of primary care medical practice. In S. Chaiklin & J. Lave (Eds.), *Understanding practice: Perspectives on activity and context* (pp. 64–103). Cambridge University Press.

Engeström, Y. (1996b). Developmental work research as educational research: Looking ten years back and into the zone of proximal development. *Nordisk Pedagogik, 16*(3), 131–143.

Engeström, Y. (1999a). Activity theory and individual and social transformation. In Y. Engeström, R. Miettinen, & R.-L. Punamaki (Eds.), *Perspectives on activity theory* (pp. 19–38). Cambridge University Press.

Engeström, Y. (1999b). Expansive visibilization of work: An activity-theoretical perspective. *Computer Supported Cooperative Work, 8*(1–2), 63–93.

Engeström, Y. (2000). Activity theory as a framework for analyzing and redesigning work. *Ergonomics, 43*(7), 960–974. https://doi.org/10.1080/001401300409143

Engeström, Y. (2001). Expansive learning at work: Toward an activity theoretical reconceptualization. *Journal of Education and Work, 14*(1), 133–156. https://doi.org/10.1080/13639080020028747

Engeström, Y. (2004). New forms of learning in co-configuration work. *Journal of Workplace Learning, 16*(1–2), 11–21.

Engeström, Y. (2005). Knotworking to create collaborative intentionality capital in fluid organizational fields. In M. M. Beyerlein, S. T. Beyerlein, & F. A. Kennedy (Eds.), *Collaborative capital: Creating intangible value* (Vol. 11, pp. 307–336). Elsevier. https://doi.org/10.1016/S1572-0977(05)11011-5

Engeström, Y. (2006). From well-bounded ethnographies to intervening in mycorrhizae activities. *Organization Studies, 27*(12), 1783–1793. https://doi.org/10.1177/0170840606071898

Engeström, Y. (2007). Enriching the theory of expansive learning: Lessons from journeys toward coconfiguration. *Mind, Culture, and Activity, 14*(1–2), 23–39.

Engeström, Y. (2008). *From teams to knots: Studies of collaboration and learning at work*. Cambridge University Press.

Engeström, Y. (2009a). The future of activity theory: A rough draft. In A. Sannino, H. Daniels, & K. Gutierrez (Eds.), *Learning and expanding with activity theory* (pp. 303–328). Cambridge.

Engeström, Y. (2009b). Wildfire activities: New patterns of mobility and learning. *International Journal of Mobile and Blended Learning, 1*(2), 1–18.

Engeström, Y. (2011). From design experiments to formative interventions. *Theory & Psychology, 21*(5), 598–628. https://doi.org/10.1177/0959354311419252

Engeström, Y. (2014). *Learning by expanding* (2nd ed.). Cambridge University Press. (Original work published 1987)

Engeström, Y. (2016). *Studies in expansive learning: Learning what is not yet there*. Cambridge University Press.

Engeström, Y. (2018). *Expertise in transition: Expansive learning in medical work*. Cambridge University Press.

Engeström, Y. (2020). Ascending from the abstract to the concrete as a principle of expansive learning. *Psychological Science and Education, 25*(5), 31–43. https://psyjournals.ru/en/journals/pse/archive/2020_n5/Engestrem

Engeström, Y. (2023). Writing for stabilization and writing for possibility: The dialectics of representation in everyday work with vulnerable clients. In P. Rogers, D. R. Russell, J. Marine, P. Carlino, & T. Donahue (Eds.), *Writing as human activity: Implications of applications of work of Charles Bazerman* (pp. 293–314). WAC Clearinghouse.

Engeström, Y., Engeström, R., & Karkkainen, M. (1995). Polycontextuality and boundary crossing in expert cognition: Learning and problem solving in complex work activities. *Learning and Instruction, 5*(4), 319–336. https://doi.org/10.1016/0959-4752(95)00021-6

Engeström, Y., Engeström, R., & Vähääho, T. (1999). When the center does not hold: The importance of knotworking. In S. Chaiklin, M. Hedegaard, & U. J. Jensen (Eds.), *Activity theory and social practice* (pp. 345–374). Aarhus University Press.

Engeström, Y., & Escalante, V. (1996). Mundane tool or object of affection? The rise and fall of the postal buddy. In B. Nardi (Ed.), *Context and consciousness: Activity theory and human-computer interaction* (pp. 325–374). MIT Press.

Engeström, Y., & Glăveanu, V. (2012). On third generation activity theory: Interview with Yrjö Engeström. *Europe's Journal of Psychology, 8*(4), 515–518. https://doi.org/10.5964/ejop.v8i4.555

Engeström, Y., Kaatrakoski, H., Kaiponen, P., Lahikainen, J., Laitinen, A., Myllys, H., Rantavuori, J., & Sinikara, K. (2012). Knotworking in academic libraries: Two case studies from the university of Helsinki. *LIBER Quarterly, 21*(3–4), 387–405. https://doi.org/10.18352/lq.8032

Engeström, Y., Kajamaa, A., Kerosuo, H., & Laurila, P. (2010). Process enhancement versus community building: Transcending the dichotomy through expansive learning. In K. Yamazumi (Ed.), *Activity theory and fostering learning: Developmental interventions in education and work* (pp. 1–28). Center for Human Activity Theory of the University of Kansai.

Engeström, Y., & Middleton, D. (1996). *Cognition and communication at work.* Cambridge University Press.

Engeström, Y., Nummijoki, J., & Sannino, A. (2012). Embodied germ cell at work: Building an expansive concept of physical mobility in home care. *Mind, Culture, and Activity, 19*(3), 287–309. https://doi.org/10.1080/10749039.2012.688177

Engeström, Y., Pasanen, A., Toiviainen, H., & Haavisto, V. (2006). Expansive learning as collaborative concept formation at work. In K. Yamazumi, Y. Engeström, & H. Daniels (Eds.), *New learning challenges: Going beyond the industrial age system of school and work* (pp. 47–77). Kansai University Press.

Engeström, Y., Puonti, A., & Seppänen, L. (2003). Spatial and temporal expansion of the object as a challenge for reorganizing work. In D. Nicolini, S. Gherardi, & D. Yanow (Eds.), *Knowing in organizations: A practice-based approach* (pp. 151–186). Sharpe.

Engeström, Y., & Sannino, A. (2010). Studies of expansive learning: Foundations, findings and future challenges. *Educational Research Review, 5*(1), 1–24. https://doi.org/10.1016/j.edurev.2009.12.002

Engeström, Y., & Sannino, A. (2016). Expansive learning on the move: Insights from ongoing research [El aprendizaje expansivo en movimiento: Aportaciones de la investigación en curso]. *Infancia y Aprendizaje, 39*(3), 401–435. https://doi.org/10.1080/02103702.2016.1189119

Engeström, Y., & Sannino, A. (2021). From mediated actions to heterogenous coalitions: Four generations of activity-theoretical studies of work and learning. *Mind Culture and Activity, 28*(1), 4–23. https://doi.org/10.1080/10749039.2020.1806328

Engeström, Y., Sannino, A., Bal, A., Lotz-Sisitka, H., Pesanayi, T., Chikunda, C., Lesama, M. F., Picinatto, A. C., Querol, M. P., & Lee, Y.-J. (2016). Agentive learning for sustainability and equity: Communities, cooperatives and social movements as emerging foci of the learning sciences. *ICLS 2016 Proceedings*, 1048–1054.

Escobar, A. (2017). *Designs for the pluriverse: Radical interdependence, autonomy, and the making of worlds.* Duke University Press.

Evans-Pritchard, E. E. (1940). *The Nuer.* Clarendon Press; Oxford University Press.

Ewing, E. T. (2001). Restoring teachers to their rights: Soviet education and the 1936 denunciation of pedology. *History of Education Quarterly, 41*(4), 471–493.

Faigley, L. (1986). Competing theories of process: A critique and a proposal. *College English, 48*(6), 527–542. https://doi.org/10.2307/376707

Fitzpatrick, S. (1994). *The Russian Revolution.* Oxford University Press.

Flower, L. (1979). Writer-based prose: A cognitive basis for problems in writing. *College English, 41*(1), 19–37. https://doi.org/10.2307/376357

Flower, L. (1994). *The construction of negotiated meaning: A social cognitive theory of writing*. Southern Illinois University Press.

Foot, K. A. (2002). Pursuing an evolving object: A case study in object formation and identification. *Mind, Culture, and Activity*, *9*(2), 132–149.

Fraiberg, S. (2017). Start-up nation: Studying transnational entrepreneurial practices in Israel's start-up ecosystem. *Journal of Business and Technical Communication*, *31*(3), 350–388. https://doi.org/10.1177/1050651917695541

Fried, M. H. (1975). *The notion of tribe*. Cummings.

Furstenau, B. (2003). Exploration of an industrial enterprise as a method of boundary crossing in vocational education. In T. Tuomi-Gröhn & Y. Engeström (Eds.), *Between school and work: New perspectives on transfer and boundary-crossing* (pp. 85–118). Pergamon.

Garcia, L. N. (2021). Soviet psychology and its utopias: Historical reflections for current science. In A. Yasnitsky (Ed.), *A history of Marxist psychology: The golden age of Soviet science* (pp. 193–209). Routledge.

Geertz, C. (1983). *Local knowledge: Further essays in interpretive anthropology*. Basic Books.

Geisler, C. (1991). Toward a sociocognitive model of literacy: Constructing mental models in a philosophical conversation. In C. Bazerman & J. Paradis (Eds.), *Textual dynamics of the professions: Historical and contemporary studies of writing in professional communities*, 171–190.

Geisler, C., Bazerman, C., Doheny-Farina, S., Gurak, L., Haas, C., Johnson-Eilola, J., Kaufer, D. S., Lunsford, A., Miller, C. R., Winsor, D. A., & Yates, J. (2001). IText: Further directions for research on the relationship between information technology and writing. *Journal of Business and Technical Communication*, *15*(3), 269–308.

González, V. M., Nardi, B., & Mark, G. (2009). Ensembles: Understanding the instantiation of activities. *Information Technology and People*, *22*(2), 109–131. https://doi.org/10.1108/09593840910962195

Granott, N. (1998). Unit of analysis in transit: From the individual's knowledge to the ensemble process. *Mind, Culture, and Activity*, *5*(1), 42–66.

Griffiths, D., Boisot, M., & Mole, V. (1998). Strategies for managing knowledge assets: A tale of two companies. *Technovation*, *18*(8–9), 529–588.

Guba, E. G., & Lincoln, Y. S. (1994). Competing paradigms in qualitative research. *Handbook of Qualitative Research*, *2*(163–194), 105.

Guile, D. (2010). *The learning challenge of the knowledge economy*. Sense Publishers.

Gygi, K., & Zachry, M. (2010). Productive tensions and the regulatory work of genres in the development of an engineering communication workshop in a transnational corporation. *Journal of Business and Technical Communication*, *24*(3), 358–381.

Haapasaari, A., Engeström, Y., & Kerosuo, H. (2016). The emergence of learners' transformative agency in a change laboratory intervention. *Journal of Education and Work*, *29*(2), 232–262. https://doi.org/10.1080/13639080.2014.900168

Haas, C. (1996). *Writing technology: Studies on the materiality of literacy*. Lawrence Erlbaum Associates.

Haas, C. (1999). On the relationship between old and new technologies. *Computers and Composition, 16*(2), 209–228.

Haenen, J. (1996). *Piotr Gal'perin: Psychologist in Vygotsky's footsteps*. Nova Science Publishers.

Hairston, M. (1985). Breaking our bonds and reaffirming our connections. *College Composition and Communication, 36*(3): 272–282.

Harrower, M. (1984). *Kurt Koffka: An unwitting self-portrait*. University Press of Florida.

Hashimov, E., & McNely, B. (2012). Left to their own devices: Ad hoc genres and the design of transmedia narratives. *Proceedings of the 30th ACM International Conference on Design of Communication* (pp. 251–259). https://doi.org/10.1145/2379057.2379105

Holland, D., & Reeves, J. R. (1996). Activity theory and the view from somewhere: Team perspectives on the intellectual work of programming. In B. Nardi (Ed.), *Context and consciousness: Activity theory and human-computer interaction* (pp. 257–282). MIT Press.

Homskaya, E. D. (2013). *Alexander Romanovich Luria: A scientific biography*. Springer Science & Business Media.

Hopwood, N. (2022). Agency in cultural-historical activity theory: Strengthening commitment to social transformation. *Mind, Culture, and Activity, 29*(2), 108–122.

Hutchins, E. (1995). *Cognition in the wild*. MIT Press.

Hyysalo, S. (2005). Objects and motives in a product design process. *Mind, Culture, and Activity, 12*(1), 19–36.

Ilyenkov, E. V. (1977). *Dialectical logic: Essays on its history and theory*. Progress.

Ilyenkov, E. V. (1982). *The dialectics of the abstract and the concrete in Marx's Capital*. Progress.

Jerolmack, C., & Khan, S. (2017). The analytic lenses of ethnography. *Socius, 3*, 1–11.

John-Steiner, V. (2014). Rhythms of collaboration. In A. Blunden (Ed.), *Collaborative projects: An interdisciplinary study* (pp. 114–125). Brill.

Jones, P. E. (2018). Vygotsky and Marx—Resetting the relationship. In A. Yasnitsky (Ed.), *Questioning Vygotsky's legacy: Scientific psychology or heroic cult* (pp. 22–42). Routledge.

Joravsky, D. (1989). *Russian psychology: A critical history*. Blackwell.

Jornet, A., & Damşa, C. (2021). Unit of analysis from an ecological perspective: Beyond the individual/social dichotomy. *Learning, Culture and Social Interaction, 31*(B), 100329.

Jungk, R., & Mullert, N. (1988). *Future workshops: How to create desirable futures*. Institute for Social Inventions.

Kaatrakoski, H., & Lahikainen, J. (2016). "What we do every day is impossible": Managing change by developing a knotworking culture in an academic library. *Journal of Academic Librarianship, 42*(5), 515–521.

Kain, D., & Wardle, E. (2005). Building context: Using activity theory to teach about genre in multi-major professional communication courses. *Technical Communication Quarterly, 14*(2), 113–139. https://doi.org/10.1207/s15427625tcq1402_1

Kaptelinin, V. (1996). *Activity theory: Implications for human-computer interaction.* MIT Press.

Kaptelinin, V. (2005). The object of activity: Making sense of the sense-maker. *Mind, Culture, and Activity, 12*(1), 4–18.

Kaptelinin, V. (2015). Activity theory. *The encyclopedia of human-computer interaction* (2nd ed.). Interaction Design Foundation. https://www.interaction-design.org/literature/book/the-encyclopedia-of-human-computer-interaction-2nd-ed/activity-theory

Kaptelinin, V., & Nardi, B. A. (2006). *Acting with technology: Activity theory and interaction design.* MIT Press.

Kaptelinin, V., & Nardi, B. (2012). *Activity theory in HCI: Fundamentals and reflections* (Vol. 5, No. 1, p. 105). Morgan & Claypool.

Karanasios, S., & Allen, D. (2018). Activity theory in information systems research. *Information Systems Journal, 28*(3), 439–441. https://doi.org/10.1111/isj.12184

Karlsson, A.-M. (2009). Positioned by reading and writing: Literacy practices, roles, and genres in common occupations. *Written Communication, 26*(1), 53–76. https://doi.org/10.1177/0741088308327445

Karpov, Y. V., & Haywood, H. C. (1998). Two ways to elaborate Vygotsky's concept of mediation: Implications for instruction. *American Psychologist, 53*(1), 27–36. https://doi.org/10.1037/0003-066X.53.1.27

Keiler, P. (2012). 'Cultural-historical theory' and 'cultural-historical school': from myth (back) to reality. *Dubna Psychological Journal,* no. 1, 1–33.

Keiler, P. (2018). A history of the social construction of the "cultural-historical." In A. Yasnitsky (Ed.), *Questioning Vygotsky's legacy: Scientific psychology or heroic cult* (pp. 91–130). Routledge.

Kellogg, D. (2011). Untangling a genetic root of thinking and speech: Towards a textology of tool and sign in child development. *Dubna Psychological Journal, 4,* 85–97.

Kerosuo, H. (2015). BIM-based collaboration across organizational and disciplinary boundaries through knotworking. *Procedia Economics and Finance, 21*(15), 201–208. https://doi.org/10.1016/s2212-5671(15)00168-9

Kerosuo, H. (2017). Transformative agency and the development of knotworking in building design. In M. Goller & S. Paloniemi (Eds.), *Agency at work: An agentic perspective on professional learning and development* (pp. 331–349). Springer. https://doi.org/10.1007/978-3-319-60943-0

Kerosuo, H. (2018). Knotworking: An emerging form of multi-party collaboration between clients, users and designers in school design. In H. M. Tse, H. Daniels, A. Stables, & S. Cox (Eds.), *Designing buildings for the futures of schooling: Contemporary visions for education* (pp. 131–148). Routledge.

Kerosuo, H., & Toiviainen, H. (2011). Expansive learning across workplace boundaries. *International Journal of Educational Research, 50*(1), 48–54. https://doi.org/10.1016/j.ijer.2011.04.009

Kinneavy, J. L. (1971). *A theory of discourse: The aims of discourse*. Prentice-Hall.

Kirsch, G., & Sullivan, P. A. (1992). *Methods and methodology in composition research*. Southern Illinois University Press.

Köhler, W. (2013). *The mentality of apes*. Routledge. (Original work published 1917)

Korpela, J. (2015). Significance of knotworking from the client's point of view. *Procedia Economics and Finance, 21*(2007), 209–216. https://doi.org/10.1016/s2212-5671(15)00169-0

Korpela, J., & Kerosuo, H. (2014). Working together in a knot: The simultaneity and pulsation of collaboration in an early phase of building design. *Proceedings 30th Annual ARCOM Conference, September*, 865–874.

Kotik-Friedgut, B., & Friedgut, T. H. (2008). A man of his country and his time: Jewish influences on Lev Semionovich Vygotsky's world view. *History of Psychology, 11*(1), 15–39. https://doi.org/10.1037/1093-4510.11.1.15

Kozulin, A. (1984). *Psychology in Utopia: Toward a social history of Soviet psychology*. MIT Press.

Kozulin, A. (1986). Vygotsky in context. In A. Kozulin (Ed.), *Thought and Language* (pp. xi–lvi). MIT Press.

Kozulin, A. (1999). *Vygotsky's psychology: A biography of ideas*. Harvard University Press.

Kozyrev, A. V., & Turko, P. A. (2000). Professor L. S. Vygotsky's "pedological school." *Journal of Russian and East European Psychology, 38*(6), 59–74. (Original work published 1936)

Kraft, P., & Bansler, J. P. (1994). The collective resource approach: The Scandinavian experience. *Scandinavian Journal of Information Systems, 6*(1), 71–84.

Krementsov, N. (1996). *Stalinist science*. Princeton University Press.

Kuutti, K. (1996). Activity theory as a potential framework for human-computer interaction research. In B. A. Nardi (Ed.), *Context and consciousness: Activity theory and human-computer interaction* (pp. 17–44). MIT Press.

Lamdan, E., & Yasnitsky, A. (2016). Did Uzbeks have illusions? The Luria-Koffka controversy of 1932. In A. Yasnitsky & R. V. D. Veer (Eds.), *Revisionist revolution in Vygotsky studies* (pp. 175–200). Routledge.

Langemeyer, I., & Roth, W.-M. (2006). Is cultural-historical activity theory threatened to fall short of its own principles and possibilities as a dialectical social science? *Outlines. Critical Practice Studies, 8*(2), 20–42.

Latour, B. (1986). Drawing things together. In H. Kuklick (Ed.), *Knowledge and society: Studies in the sociology of culture past and present* (pp. 1–40). JAI Press.

Latour, B. (1987). *Science in action: How to follow scientists and engineers through society*. Harvard University Press.

Latour, B. (1988). *The pasteurization of France*. Harvard University Press.

Latour, B. (1996). *Aramis, or the love of technology*. Harvard University Press.

Latour, B. (1999). *Pandora's hope: Essays on the reality of science studies*. Harvard University Press.

Latour, B. (2006). *Reassembling the social: An introduction to actor-network-theory*. Oxford University Press.

Latour, B. (2010). *The making of law: An ethnography of the Conseil d'Etat*. Polity.

Latour, B. (2018). *Rejoicing: Or the torments of religious speech*. John Wiley & Sons.

Latour, B., & Woolgar, S. (1979). *Laboratory life: The social construction of scientific facts*. Sage Publications.

Lauer, J. M., & Asher, J. W. (1988). *Composition research: Empirical designs*. Oxford University Press.

Lave, J., Murtaugh, M., & de la Rocha, O. (1984). *The dialectic of arithmetic in grocery shopping*. Harvard University Press.

Law, J. (2002). *Aircraft stories: Decentering the object in technoscience*. Duke University Press.

Law, J. (2004). *After method: Mess in social science research*. Routledge.

Law, J., & Singleton, V. (2005). Object lessons. *Organization*, *12*(3), 331–331.

Lektorsky, V. A. (1999). Activity theory in a new era. In Y. Engeström, R. Miettinen, & R.-L. Punamäki (Eds.), *Perspectives on activity theory* (pp. 65–69). Cambridge University Press.

Lémonie, Y., Grosstephan, V., & Tomás, J. (2021). From a sociological given context to changing practice: Transforming problematic power relations in educational organizations to overcome social inequalities. *Frontiers in Psychology*, *11*(January), 1–16. https://doi.org/10.3389/fpsyg.2020.608502

Lenin, V. I. (1987). *Essential works of Lenin* (H. M. Christman, Ed.). Dover. (Original work published 1917)

Leont'ev, A. N. (1969). On the biological and social aspects of human development: The training of auditory ability. In M. Cole & I. Maltzman (Eds.), *A handbook of contemporary Soviet psychology* (pp. 423–440). Basic Books. (Original work published 1960)

Leont'ev, A. N. (1978). *Activity, consciousness, and personality*. Prentice-Hall. (Original work published 1975)

Leont'ev, A. N. (1981). The problem of activity in psychology. In J. V. Wertsch (Ed.), *The concept of activity in Soviet psychology* (pp. 37–71). M. E. Sharpe. (Original work published 1959)

Leont'ev, A. N., & Zaporozhets, A. V. (1960). *Rehabilitation of hand function*. Pergamon. (Original work published 1945)

Leontiev, A. A. (2005). The life and creative path of A. N. Leontiev. *Journal of Russian and East European Psychology*, *43*(3), 8–69. https://doi.org/10.1080/10610405.2005.11059249

Leontiev, A. N. (1957). The nature and formation of human psychic properties. In B. Simon (Ed.), *Psychology in the Soviet Union* (pp. 226–232). Routledge & Kegan Paul. (Original work published 1954)

Leontiev, A. N. (1961). The present tasks of Soviet psychology. In *Soviet psychology: A symposium* (pp. 31–48). Philosophical Library. (Original work published 1949)

Leontiev, A. N. (1972). Some prospective problems of Soviet psychology. In J. M. Brozek (Ed.), *Psychology in the USSR: An historical perspective* (pp. 144–157). International Arts and Sciences Press. (Original work published 1967)

Leontiev, A. N. (2005a). Study of the environment in the pedological works of LS Vygotsky: A critical study. *Journal of Russian and East European Psychology*, *43*(4), 8–28. https://doi.org/10.1080/10610405.2005.11059254 (Original work written 1937)

Leontiev, A. N. (2005b). The fundamental processes of mental life. *Journal of Russian and East European Psychology*, *43*(4), 72–75. (Original work published 1940)

Leontiev, A. N. (2005c). The genesis of activity. *Journal of Russian and East European Psychology*, *43*(4), 58–71. (Original work published 1940)

Leontiev, A. N., & Luria, A. R. (2005). The problem of the development of the intellect and learning in human psychology. *Journal of Russian and East European Psychology*, *43*(4), 34–47. (Original work published 1937)

Leontyev, A. N. (1981). *Problems of the development of mind*. Progress. (Original work published 1959)

Leontyev, A. N. (2005). Will. *Journal of Russian and East European Psychology*, *43*(4), 76–92. (Original work published 1978)

Leontyev, A. N. (2009a). *Activity and consciousness*. Marxists Internet Archive, https://www.marxists.org/archive/leontev/works/activity-consciousness.pdf. (Original work published 1977)

Leontyev, A. N. (2009b). *The development of mind*. Marxists Internet Archive, https://www.marxists.org/admin/books/activity-theory/leontyev/development-mind.pdf. (Original work published 1959)

Levant, A. (2018). Two, three, many strands of activity theory! *Educational Review*, *70*(1), 100–108. https://doi.org/10.1080/00131911.2018.1388619

Levitin, K. (1982). *One is not born a personality: Profiles of Soviet educational psychologists*. Progress.

Lewin, K. (1935). *A dynamic theory of personality*. McGraw-Hill.

Little, J. (2007). Achieving objectivity through genred activity: A case study. *Journal of Technical Writing and Communication*, *37*(1), 75–94. https://doi.org/10.2190/T85G-0265-P628-6236

Lompscher, J. (2006). The cultural-historical development of activity theory: Some aspects of development. In P. Sawchuk, N. Duarte, & M. Elhammoumi (Eds.), *Critical perspectives on activity: Explorations across education, work, and everyday life* (pp. 35–51). Cambridge University Press.

Long, K. J. (2004). Unit of analysis. In M. S. Lewis-Beck, A. Bryman, & T. F. Lao (Eds.), *The SAGE encyclopedia of social science research methods* (p. 1158). SAGE Publications.

Lotz-Sisitka, H., Mukute, M., Chikunda, C., Baloi, A., & Pesanayi, T. (2017). Transgressing the norm: Transformative agency in community-based learning for sustainability in southern African contexts. *International Review of Education, 63*(6), 897–914. https://doi.org/10.1007/s11159-017-9689-3

Ludvigsen, S. R., Havnes, A., & Lahn, L. C. (2003). Workplace learning across activity systems: A case study of sales engineers. In T. Tuomi-Gröhn & Y. Engeström (Eds.), *Between school and work: New perspectives on transfer and boundary-crossing* (pp. 291–310). Pergamon.

Luria, A. R. (1928). The problem of the cultural behavior of the child. *Pedagogical Seminary and Journal of Genetic Psychology, 35*(4), 493–506.

Luria, A. R. (1934). The second psychological expedition to Central Asia. *Pedagogical Seminary and Journal of Genetic Psychology, 44*(1), 255–259. https://doi.org/10.1080/08856559.1934.10532497

Luria, A. R. (1957a). Psychopathological research in the U.S.S.R. In B. Simon (Ed.), *Psychology in the Soviet Union* (pp. 279–287). Routledge & Kegan Paul.

Luria, A. R. (1957b). The role of language in the formation of temporary connections. In B. Simon (Ed.), *Psychology in the Soviet Union* (pp. 115–129). Stanford University Press.

Luria, A. R. (1960). *The nature of human conflicts*. Grove Press.

Luria, A. R. (1966). *Higher cortical functions in man*. Basic Books.

Luria, A. R. (1968). *Mind of a mnemonist: A little book about a vast memory*. Basic Books.

Luria, A. R. (1972). *The man with a shattered world: The history of a brain wound*. Basic Books.

Luria, A. R. (1973). *The working brain: An introduction to neuropsychology*. Basic Books.

Luria, A. R. (1976). *Cognitive development: Its cultural and social foundations*. Harvard University Press.

Luria, A. R., Cole, M., & Cole, S. (1979). *The making of mind: A personal account of Soviet psychology*. Harvard University Press.

Luria, A. R., & Udovich, F. I. A. (1959). *Speech and the development of mental processes in the child; an experimental investigation*. Staples Press. (Original work published 1956)

MacKenzie, D. (2008). *Material markets: How economic agents are constructed*. Oxford University Press.

Mahn, H. (2010). Vygotsky's methodological approach. In A. Toomela & J. Valsiner (Eds.), *Methodological thinking in psychology: 60 years gone astray?* (pp. 297–324). Information Age Publishing.

Maidansky, A. (2018). Spinoza in cultural-historical psychology. *Mind, Culture, and Activity, 25*(4), 355–364.

Maire, L. F. I., & LaMothe, J. D. (1975). *Soviet and Czechslovakian parapsychology research*. Defense Intelligence Agency.

Maistre, C. L., & Paré, A. (2004). Learning in two communities: The challenge for universities and workplaces. *Journal of Workplace Learning, 16*(1–2), 44–52. https://doi.org/10.1108/13665620410521503

Marx, K. (1845). *Theses on Feuerbach*. Marxists Internet Archive. https://www.marxists.org/archive/marx/works/1845/theses/theses.htm

Marx, K. (1922). *Critique of the Gotha Program*. Socialist Labor Party. (Original work published 1875)

Marx, K. (1990). *Capital* (Vol. 1). Penguin Books.

Maslina, M. N. (1950). The Leontiev critique: For Bolshevist partisanship in questions of psychology. In J. Wortis (Ed.), *Soviet psychiatry* (pp. 295–300). Williams and Wilkins.

Matusov, E. (2007). In search of "the appropriate" unit of analysis for sociocultural research. *Culture and Psychology, 13*(3), 307–333. https://doi.org/10.1177/1354067X07079887

Matusov, E. (2013). *Journey into dialogic pedagogy*. Nova Science Publishers.

Matusov, E., & St. Julien, J. (2004). Print literacy as oppression: Cases of bureaucratic, colonial, and totalitarian literacies and their implications for schooling. *Text, 24*(2), 197–244. https://doi.org/10.1515/text.2004.008

McCarthy, J. E., Grabill, J. T., Hart-Davidson, W., & McLeod, M. (2011). Content management in the workplace: Community, context, and a new way to organize writing. *Journal of Business and Technical Communication, 25*(4), 367–395.

McCarthy, L. P. (1991). A psychiatrist using DSM-III. In C. Bazerman, J. G. Paradis, & J. Paradis (Eds.), *Textual dynamics of the professions: Historical and contemporary studies of writing in professional communities* (pp. 358–381). University of Wisconsin Press.

McLeish, J. (1975). *Soviet psychology: History, theory, context*. Methuen & Co.

McNair, L. D., & Paretti, M. C. (2010). Activity theory, speech acts, and the "doctrine of infelicity": Connecting language and technology in globally networked learning environments. *Journal of Business and Technical Communication, 24*(3), 323–357.

McNely, B. (2017). Moments and metagenres: Coordinating complex, multigenre narratives. *Journal of Business and Technical Communication, 31*(4), 443–480.

Merriam, S. B., & Tisdell, E. J. (2015). *Qualitative research: A guide to design and implementation*. John Wiley & Sons.

Metraux, A. (2021). Alexander Luria: Marxist psychologist and transnational scientific broker: A personal account. In A. Yasnitsky (Ed.), *A history of Marxist psychology: The golden age of Soviet science* (pp. 156–190). Routledge.

Miettinen, R. (1998). Object construction and networks in research work: The case of research on cellulose-degrading enzymes. *Social Studies of Science, 28*(3), 423–463.

Miettinen, R. (1999). The riddle of things: Activity theory and actor-network theory as approaches to studying innovations. *Mind, Culture, and Activity*, *6*(3), 170–195.

Miettinen, R. (2005). Object of activity and individual motivation. *Mind, Culture, and Activity*, *12*(1), 52–69.

Miettinen, R. (2008). Contradictions of high-technology capitalism and the emergence of new forms of work. In A. Sannino, H. Daniels, & K. D. Gutierrez (Eds.), *Learning and expanding with activity theory* (pp. 160–175). Cambridge University Press.

Miettinen, R., & Hasu, M. (2002). Articulating user needs in collaborative design: Towards an activity-theoretical approach. *Computer Supported Cooperative Work*, *11*(1–2), 129–151.

Miettinen, R., & Virkkunen, J. (2005). Epistemic objects, artefacts and organizational change. *Organization*, *12*(3), 437–456.

Miles, M. B., Huberman, A. M., & Saldaña, J. (2020). *Qualitative data analysis: A methods sourcebook* (4th ed.). SAGE Publications.

Miller, C. R. (1984). Genre as social action. *Quarterly Journal of Speech*, *70*(2), 151–167.

Miller, R. (2011). *Vygotsky in perspective*. Cambridge.

Mol, A. (2002). *The body multiple: Ontology in medical practice*. Duke University Press.

Mudokwani, K., & Mukute, M. (2019). Exploring group solidarity for insights into qualities of T-learning. *Sustainability (Switzerland)*, *11*(23), 1–11. https://doi.org/10.3390/su11236825

Mukute, M. (2016). Dialectical critical realism and cultural historical activity theory (CHAT): Exploring and expanding learning processes in sustainable agriculture workplace contexts. In L. Price & H. Lotz-Sisitka (Eds.), *Critical realism, environmental learning and social-ecological change* (pp. 190–211). Routledge.

Mukute, M., & Lotz-Sisitka, H. (2012). Working with cultural-historical activity theory and critical realism to investigate and expand farmer learning in Southern Africa. *Mind, Culture, and Activity*, *19*(4), 342–367. https://doi.org/10.1080/10749039.2012.656173

Mukute, M., Mudokwani, K., McAllister, G., & Nyikahadzoi, K. (2018). Exploring the potential of developmental work research and change laboratory to support sustainability transformations: A case study of organic agriculture in Zimbabwe. *Mind, Culture, and Activity*, *25*(3), 229–246. https://doi.org/10.1080/10749039.2018.1451542

Nardi, B. A. (1996a). Activity theory and human-computer interaction. In B. A. Nardi (Ed.), *Context and consciousness: Activity theory and human-computer interaction* (pp. 7–16). MIT Press.

Nardi, B. A. (Ed.). (1996b). *Context and consciousness: Activity theory and human-computer interaction*. MIT Press.

Nardi, B. A. (2005). Objects of desire: Power and passion in collaborative activity. *Mind, Culture, and Activity*, *12*(1), 37–51.

Nardi, B. A. (2007). Placeless organizations: Collaborating for transformation. *Mind, Culture, and Activity*, *14*(1), 5–22.

Nardi, B. A. (2010). *My life as a night elf priest: An anthropological account of World of Warcraft*. University of Michigan Press.

North, S. M. (1987). *The making of knowledge in composition: Portrait of an emerging field*. Boynton/Cook Publishers.

Nummijoki, J., & Engeström, Y. (2013). Towards co-configuration in home care of the elderly: Cultivating agency by designing and implementing the mobility agreement. In H. Daniels, A. Edwards, Y. Engeström, & S. R. Ludvigsen (Eds.), *Activity theory in practice: Promoting learning across boundaries and agencies* (pp. 49–71). Routledge. https://doi.org/10.4324/9780203609439

Nummijoki, J., Engeström, Y., & Sannino, A. (2018). Defensive and expansive cycles of learning: A study of home care encounters. *Journal of the Learning Sciences, 27*(2), 224–264. https://doi.org/10.1080/10508406.2017.1412970

Nygaard, K. (1992). *How many choices do we make? How many are difficult?* Springer-Verlag.

Nystrand, M., Greene, S., & Wiemelt, J. (1993). Where did composition studies come from? An intellectual history. *Written Communication, 10*(3), 267–333.

Odell, L., & Goswami, D. (1985). *Writing in nonacademic settings*. Guilford Press.

Olson, D. R. (2016). *The mind on paper: Reading, consciousness and rationality*. Cambridge University Press.

Opel, D. S., & Hart-Davidson, W. (2019). The primary care clinic as writing space. *Written Communication, 36*(3), 348–378. https://doi.org/10.1177/0741088319839968

Osipov, G. V. (2009). The rebirth of sociology in Russia. *Russian Social Science Review, 50*(6), 80–108. https://doi.org/10.1080/10611428.2009.11065377

Payne, T. R. (1968). *S.L. Rubinštejn and the philosophical foundations of Soviet psychology* (Vol. 30). D. Reidel Publishing Company.

Peim, N. (2009). Activity theory and ontology. *Educational Review, 61*(2), 167–180. https://doi.org/10.1080/00131910902846874

Petrovsky, A. (1990). *Psychology in the Soviet Union: A historical outline*. Progress.

Ploettner, J., & Tresseras, E. (2016). An interview with Yrjö Engeström and Annalisa Sannino on activity theory. *Bellaterra Journal of Teaching & Learning Language & Literature, 9*(4), 87–98.

Popham, S. L. (2005). Forms as boundary genres in medicine, science, and business. *Journal of Business and Technical Communication, 19*(3), 279–303.

Prior, P., & Schaffner, S. (2011). Bird identification as a family of activities: Motives, mediating artifacts, and laminated assemblages. *Ethos, 39*(1), 51–70. https://doi.org/10.1111/j.1548-1352.2010.01170.x

Proctor, H. (2020). *Psychologies in revolution: Alexander Luria's' romantic science' and Soviet social history*. Springer.

Prokopis, D., Sannino, A., & Mykkänen, A. (2022). Toward a new beginning: Exploring the instructional dynamics of expansive learning with workers in a youth supported housing unit. *Journal of Workplace Learning, 34*(7), 628–642.

Propen, A. D., & Schuster, M. L. (2010). Understanding genre through the lens of advocacy: The rhetorical work of the victim impact statement. *Written Communication, 27*(1), 3–35. https://doi.org/10.1177/0741088309351479

Rabinow, P. (2007). *Reflections on fieldwork in Morocco*. University of California Press.

Radzikhovskii, L. (2020). Reminiscence about future Marxist psychology: One hundred years of solitude. In A. Yasnitsky (Ed.), *A history of Marxist psychology: The golden age of Soviet science* (pp. 25–57). Routledge.

Rahmani, L. (1973). *Soviet psychology: Philosophical, theoretical, and experimental issues*. International Universities Press.

Razmyslov, P. (2000). On Vygotsky's and Luria's "cultural-historical theory of psychology." *Journal of Russian and East European Psychology, 38*(6), 45–58. (Original work published 1934)

Rogers, E. M. (2003). *Diffusion of innovations* (5th ed.). Free Press.

Roller, M. R., & Lavrakas, P. J. (2015). *Applied qualitative research design: A total quality framework approach*. Guilford Publications.

Rosenthal, B. G. (2002). *New myth, new world: From Nietzsche to Stalinism*. Penn State University Press.

Ruckriem, G. (2009). Digital technology and mediation: A challenge to activity theory. In A. Sannino, H. Daniels, & K. D. Gutierrez (Eds.), *Learning and expanding with activity theory* (pp. 88–111). Cambridge University Press.

Rudneva, E. I. (2000). Vygotsky's pedological distortions. *Journal of Russian and East European Psychology, 38*(6), 75–94. (Original work published 1937)

Russell, D. R. (1991). *Writing in the academic disciplines, 1870–1990: A curricular history*. Southern Illinois University Press.

Russell, D. R. (1993). Vygotsky, Dewey, and externalism: Beyond the student/discipline dichotomy. *Journal of Advanced Composition, 13*(1), 173–197.

Russell, D. R. (1995). Activity theory and its implications for writing instruction. In J. Petraglia (Ed.), *Reconceiving writing, rethinking writing instruction* (pp. 51–78). Lawrence Erlbaum Associates.

Russell, D. R. (1997a). Rethinking genre in school and society: An activity theory analysis. *Written Communication, 14*(4), 504–554.

Russell, D. R. (1997b). Writing and genre in higher education and workplaces: A review of studies that use cultural-historical activity theory. *Mind, Culture, and Activity, 4*(4), 224–237.

Russell, D. R. (2009). Uses of activity theory in written communication research. In A. Sannino, H. Daniels, & K. D. Gutierrez (Eds.), *Learning and expanding with activity theory* (pp. 40–52). Cambridge University Press.

Russell, D. R. (2010). Writing in multiple contexts: Vygotskian CHAT meets the phenomenology of genre. In C. Bazerman, R. Krut, K. Lunsford, S. McLeod, S. Null, P. Rogers, & A. Stansell (Eds.), *Traditions of writing research* (pp. 353–364). Routledge.

Saarelma, O. (1993). Descriptions of subjective networks as a mediator of developmental dialogue. *The Quarterly Newsletter of the Laboratory of Comparative Human Cognition, 15*, 102–112.

Sacks, O. (1999). *The man who mistook his wife for a hat: And other clinical tales*. HarperCollins Publishers.

Sacks, O. (2014). Luria and "romantic science." In A. Yasnitsky, R. Van Der Veer, & M. Ferrari (Eds.), *The Cambridge handbook of cultural-historical psychology* (pp. 517–528). Cambridge University Press.

Sakharov. (1994). Methods for investigating concepts. In R. Van Der Veer & J. Valsiner (Eds.), *The Vygotsky reader* (pp. 73–98). Blackwell.

Säljö, R. (2009). Learning, theories of learning, and units of analysis in research. *Educational Psychologist, 44*(3), 202–208. https://doi.org/10.1080/00461520903029030

Sannino, A. (2011). Activity theory as an activist and interventionist theory. *Theory & Psychology, 21*(5), 571–597. https://doi.org/10.1177/0959354311417485

Sannino, A. (2018). Counteracting the stigma of homelessness: The Finnish Housing First strategy as educational work. *Educação, 41*(3), 385–392.

Sannino, A. (2020a). Enacting the utopia of eradicating homelessness: Toward a new generation of activity-theoretical studies of learning. *Studies in Continuing Education, 42*(2), 163–179. https://doi.org/10.1080/0158037X.2020.1725459

Sannino, A. (2020b). Transformative agency as warping: How collectives accomplish change amidst uncertainty. *Pedagogy, Culture and Society, 30*(1), 9–33. https://doi.org/10.1080/14681366.2020.1805493

Sannino, A. (2023). Problem identification in change laboratories: Workplace learning to eradicate homelessness. In H. Bound, A. Edwards, K. Evans, & A. Chia (Eds.), *Workplace learning for changing social and economic circumstances* (pp. 201–218). Taylor & Francis.

Sannino, A., Daniels, H., & Gutierrez, K. D. (Eds.). (2009). *Learning and expanding with activity theory*. Cambridge University Press.

Sannino, A., & Engeström, Y. (2017). Co-generation of societally impactful knowledge in change laboratories. *Management Learning, 48*(1), 80–96. https://doi.org/10.1177/1350507616671285

Sannino, A., & Engeström, Y. (2018). Valuable innovations out of nonsense? Expansive organizational learning and transformative agency in the Mann Gulch disaster and in the Finnish homelessness strategy. *Teoria e Prática Em Administração, 8*(2), 60–79. https://doi.org/10.21714/2238-104x2018v8i2s-40728

Sannino, A., & Engeström, Y. (2023). In search of an experiment: From Vygotsky to Lewin and Dembo and back to the future. In G. Marisco & L. Tateo (Eds.), *Humanity in psychology: The intellectual legacy of Pina Boggi Cavallo* (pp. 159–177). Springer.

Sannino, A., Engeström, Y., & Jokinen, E. (2021). Digital peer learning for transformative professional agency: The case of homelessness practitioners in Finland. *British Journal of Educational Technology, 52*(4), 1612–1628. https://doi.org/10.1111/bjet.13117

Sannino, A., Engeström, Y., & Lahikainen, J. (2016). The dialectics of authoring expansive learning: Tracing the long tail of a change laboratory. *Journal of Workplace Learning, 28*(4), 245–262.

Sawchuk, P., Duarte, N., & Elhammoumi, M. (Eds.). (2006). *Critical perspectives on activity: Explorations across education, work, and everyday life.* Cambridge University Press.

Sawchuk, P., & Stetsenko, A. (2008). Sociological understandings of conduct for a noncanonical activity theory: Exploring intersections and complementarities. *Mind, Culture, and Activity, 15*(4), 339–360. https://doi.org/10.1080/10749030802391427

Schmandt-Besserat, D., & Erard, M. (2008). Origins and forms of writing. In C. Bazerman (Ed.), *Handbook of research on writing: History, society, school, individual, text* (pp. 7–22). Erlbaum.

Schryer, C. F. (1993). Records as genre. *Written Communication, 10*(2), 200–234.

Schryer, C. F., Afros, E., Mian, M., Spafford, M., & Lingard, L. (2009). The trial of the expert witness: Negotiating credibility in child abuse correspondence. *Written Communication, 26*(3), 215–246.

Schryer, C. F., Lingard, L., & Spafford, M. (2007). Regularized practices: Genres, improvisation, and identity formation in health-care professions. In M. Zachry & C. Thralls (Eds.), *Communicative practices in workplaces and the professions: Cultural perspectives on the regulation of discourse and organizations* (pp. 21–44). Baywood Publishing Company.

Schryer, C. F., & Spoel, P. (2005). Genre theory, health-care discourse, and professional identity formation. *Journal of Business and Technical Communication, 19*(3), 249–278.

Schuster, M. L., & Propen, A. D. (2011). *Victim advocacy in the courtroom: Persuasive practices in domestic violence and child protection cases.* Northeastern University Press.

Scribner, S., & Cole, M. (1981). *The psychology of literacy.* Harvard University Press.

Sève, L. (2021). Where is Marx in the work and thought of Vygotsky? *Mind, Culture, and Activity, 28*(4), 293–303.

Sherlock, L. (2009). Genre, activity, and collaborative work and play in World of Warcraft: Places and problems of open systems in online gaming. *Journal of Business and Technical Communication, 23*(3), 263–293.

Smagorinsky, P. (2011). *Vygotsky and literacy research.* Sense Publishers. https://doi.org/10.1007/978-94-6091-696-0

Smart, G. (2006). *Writing the economy: Activity, genre, and technology in the world of banking.* Equinox.

Smith, A., & Prior, P. (2020). A flat CHAT perspective on transliteracies development. *Learning, Culture and Social Interaction, 24,* 1–11. https://doi.org/10.1016/j.lcsi.2019.01.001

Snow, D. A., & Anderson, L. (1993). *Down on their luck: A study of homeless street people.* University of California Press.

Spafford, M., Schryer, C., Lingard, L., & Mian, M. (2010). Accessibility and order: Crossing borders in child abuse forensic reports. *Technical Communication Quarterly, 19*(2), 118–143.

Spafford, M. M., Schryer, C. F., Mian, M., & Lingard, L. (2006). Look who's talking: Teaching and learning using the genre of medical case presentations. *Journal of Business and Technical Communication, 20*(2), 121–158. https://doi.org/10.1177/1050651905284396

Spasser, M. A. (2000). Articulating collaborative activity: Design-in-use of collaborative publishing services in the flora of North America Project. *Scandinavian Journal of Information Systems 12*, 140–172.

Spinuzzi, C. (2002a). A Scandinavian challenge, a US response: Methodological assumptions in Scandinavian and US prototyping approaches. *Proceedings of the 20th Annual International Conference on Computer Documentation*, 208–215. https://doi.org/10.1145/584955.584986

Spinuzzi, C. (2002b). Toward integrating our research scope: A sociocultural field methodology. *Journal of Business and Technical Communication, 16*(1), 3–32.

Spinuzzi, C. (2003a). Compound mediation in software development: Using genre ecologies to study textual artifacts. In C. Bazerman & D. R. Russell (Eds.), *Writing selves/writing societies: Research from activity perspectives* (pp. 97–124). WAC Clearinghouse; *Mind, Culture, and Activity*.

Spinuzzi, C. (2003b). *Tracing genres through organizations: A sociocultural approach to information design*. MIT Press.

Spinuzzi, C. (2004). Four ways to investigate assemblages of texts: Genre sets, systems, repertoires, and ecologies. *SIGDOC '04: Proceedings of the 22nd Annual International Conference on Design of Communication*, 110–116.

Spinuzzi, C. (2005a). Lost in the translation: Shifting claims in the migration of a research technique. *Technical Communication Quarterly, 14*(4), 411–446.

Spinuzzi, C. (2005b). The methodology of participatory design. *Technical Communication, 52*(2), 163–174.

Spinuzzi, C. (2007). Guest editor's introduction: Technical communication in the age of distributed work. *Technical Communication Quarterly, 16*(3), 265–277.

Spinuzzi, C. (2008). *Network: Theorizing knowledge work in telecommunications*. Cambridge University Press.

Spinuzzi, C. (2010). Secret sauce and snake oil: Writing monthly reports in a highly contingent environment. *Written Communication, 27*(4), 363–409.

Spinuzzi, C. (2011). Losing by expanding: Corralling the runaway object. *Journal of Business and Technical Communication, 25*(4), 449–486.

Spinuzzi, C. (2015a). *All edge: Inside the new workplace networks*. University of Chicago Press.

Spinuzzi, C. (2015b). Toward a typology of activities: Understanding internal contradictions in multiperspectival activities. *Journal of Business and Technical Communication, 29*(1), 3–35. https://doi.org/10.1177/1050651914548277

Spinuzzi, C. (2018). *Topsight 2.0: A guide to studying, diagnosing, and fixing information flow in organizations*. Urso Press.

Spinuzzi, C. (2021a). Scaling change labs: A response to "From mediated actions to heterogenous coalitions: Four generations of activity-theoretical studies of work and learning." *Mind Culture and Activity, 28*(1), 24–31. https://doi.org/10.1080/10749039.2020.1840594

Spinuzzi, C. (2021b). What's wrong with 3GAT? In C. Donahue, K. Blewett, & C. Monroe (Eds.), *The expanding universe of writing studies: Higher education writing research* (pp. 279–296). Peter Lang.

Spinuzzi, C. (2023a). Two paths diverge in a field: Dialectics and dialogics in rhetorical genre studies. In P. Rogers, D. Russell, J. Marine, P. Carlino, & T. Donahue (Eds.), *Writing as human activity: Implications of applications of work of Charles Bazerman* (pp. 265–292). WAC Clearinghouse.

Spinuzzi, C. (2023b). What is a workplace? Principles for bounding case studies of genres, processes, objects, and organizations. *Written Communication, 40*(4), 1027–1069. https://doi.org/10.1177/07410883231185875

Spinuzzi, C., Altounian, D., Pogue, G., Cochran, R., & Zhu, L. (2018). Articulating problems and markets: A translation analysis of entrepreneurs' emergent value propositions. *Written Communication, 35*(4), 379–410. https://doi.org/10.1177/0741088318786235

Spinuzzi, C., & Guile, D. (2019). Fourth-generation activity theory: An integrative literature review and implications for professional communication. In C. Digmayer & J. Karreman (Eds.), *Proceedings of IEEE Professional Communication Society international professional communication conference* (pp. 37–45). ACM. https://doi.org/10.1109/procomm.2019.00012

Spinuzzi, C., Nelson, S., Thomson, K. S., Lorenzini, F., French, R. A., Pogue, G., & London, N. (2016). How magnets attract and repel: Interessement in a technology commercialization competition. *Written Communication, 33*(1), 3–41. https://doi.org/10.1177/0741088315614566

Stalin, J. (1939). *Foundations of Leninism* (p. 116). International Publishers. (Original work published 1924)

Star, S. L., & Griesemer, J. R. (1989). Institutional ecology, "translations" and boundary objects: Amateurs and professionals in Berkeley's Museum of Vertebrate Zoology, 1907–1939. *Social Studies of Science, 19*(3), 387–420.

Stetsenko, A. (2017). *The transformative mind: Expanding Vygotsky's approach to development and education*. Cambridge University Press.

Stetsenko, A. (2021). Scholarship in the context of a historic socioeconomic and political turmoil: Reassessing and taking stock of CHAT. Commentary on Y. Engeström and A. Sannino "from mediated actions to heterogenous coalitions: Four generations of activity-theoretical studies of work and learning." *Mind, Culture, and Activity, 28*(1), 32–43. https://doi.org/10.1080/10749039.2021.1874419

Sun, H. (2012). *Cross-cultural technology design: Crafting culture-sensitive technology for local users*. Oxford University Press.

Swarts, J. (2011). Technological literacy as network building. *Technical Communication Quarterly, 20*(3), 274–302.

Toassa, G. (2015). Is there a "Vygotskian Materialism"? Ontological and epistemological concerns for a contemporary Marxist psychology (Part II). *Dubna Psychological Journal, 3*, 66–80.

Toassa, G. (2019). Leontiev about matter and consciousness: His critique of Vygotsky in the Soviet context. *Psicologia: Teoria e Pesquisa, 35*, 1–10.

Toffler, A. (1980). *The third wave*. Bantam Books.

Toffler, A. (1990). *Powershift*. Bantam Books.

Tomasello, M. (1999). *The cultural origins of human cognition*. Harvard University Press.

Tomasello, M. (2019). *Becoming human: A theory of ontogeny*. Belknap Press of Harvard University Press.

Toomela, A. (2000). Activity theory is a dead end for cultural-historical psychology. *Culture & Psychology, 6*(3), 353–364.

Toomela, A. (2008). Commentary: Vygotskian cultural-historical and sociocultural approaches represent two levels of analysis: Complementarity instead of opposition. *Culture & Psychology, 14*(1), 57–69.

Toomela, A. (2016). What are higher psychological functions? *Integrative Psychological and Behavioral Science, 50*(1), 91–121. https://doi.org/10.1007/s12124-015-9328-0

Trotsky, L. (2005). *Literature and revolution*. Haymarket Books. (Original work published 1923)

Tulviste, P. (1991). *The cultural-historical development of verbal thinking* (M. J. T. Hall, Trans.). Nova Science Publishers.

Tuomi-Gröhn, T., & Engeström, Y. (Eds.). (2003). *Between school and work: New perspectives on transfer and boundary crossing*. Pergamon.

Tuomi-Gröhn, T., Engeström, Y., & Young, M. (2003). From transfer to boundary-crossing between school and work as a tool for developing vocational education: An introduction. In T. Tuomi-Gröhn & Y. Engeström (Eds.), *Between school and work: New perspectives on transfer and boundary-crossing* (pp. 1–15). Pergamon.

Valsiner, J. (1988). *Developmental psychology in the Soviet Union*. Indiana University Press.

Valsiner, J., & Van der Veer, R. (2000). *The social mind: Construction of the idea*. Cambridge University Press.

Valsiner, J., & Van der Veer, R. (2014). Encountering the border: Vygotsky's zona blizhaishego razvita and its implications for theories of development. In A. Yasnitsky, R. van der Veer, & M. Ferrari (Eds.), *The Cambridge handbook of cultural-historical psychology* (pp. 148–174). Cambridge University Press.

Van der Veer, R. (2000). Editor's introduction: Criticizing Vygotsky. *Journal of Russian and East European Psychology, 38*(6), 3–9.

Van der Veer, R. (2008). Multiple readings of Vygotsky. In B. van Oers, W. Wardekker, E. Elbers, & R. van der Veer (Eds.) *The transformation of learning: Advances in cultural-historical activity theory*. https://doi.org/10.1017/CBO9780511499937.004

Van der Veer, R. (2014). *Lev Vygotsky*. Bloomsbury.

Van der Veer, R., & Valsiner, J. (1991). *Understanding Vygotsky: A quest for synthesis*. Blackwell Publishers.

Van der Veer, R., & Yasnitsky, A. (2016a). Translating Vygotsky: Some problems of transnational Vygotskian science. In A. Yasnitsky & R. Van der Veer (Eds.), *Revisionist revolution in Vygotsky studies* (pp. 143–173). Routledge.

Van Der Veer, R., & Yasnitsky, A. (2016b). Vygotsky the published: Who wrote Vygotsky and what Vygotsky actually wrote. In A. Yasnitsky & R. V. D. Veer (Eds.), *Revisionist revolution in Vygotsky studies* (pp. 73–93). Routledge.

Van Maanen, J. (2011). *Tales of the field: On writing Ethnography* (2nd ed.). University of Chicago Press.

Varpio, L., Hall, P., Lingard, L., & Schryer, C. F. (2008). Interprofessional communication and medical error: A reframing of research questions and approaches. *Academic Medicine, 83*(10), S76–S81.

Varpio, L., Spafford, M. M., Schryer, C. F., & Lingard, L. (2007). Seeing and listening: A visual and social analysis of optometric record-keeping practices. *Journal of Business and Technical Communication, 21*(4), 343–375.

Veresov, N. (2010). Forgotten methodology: Vygotsky's case. In A. Toomela & J. Valsiner (Eds.), *Methodological thinking in psychology: 60 years gone astray?* (pp. 267–295). Information Age Publishing.

Victor, B., & Boynton, A. C. (1998). *Invented here: Maximizing your organization's internal growth and profitability*. Harvard Business School Press.

Vogel, C., Mukute, M., Coetzer, K., & Gwata, M. (2021). Creating a climate of change in the City of Johannesburg: Co-learning to adapt to climate change. *South African Journal of Science, 117*(9–10). https://doi.org/10.17159/sajs.2021/7929

Vygotski, L. S. (1929). II. The problem of the cultural development of the child. *The Pedagogical Seminary and Journal of Genetic Psychology, 36*(3), 415–434.

Vygotsky, L. S. (1927). *The historical meaning of the crisis in psychology: A methodological investigation*. Marxists Internet Archive. https://www.marxists.org/archive/vygotsky/works/crisis/

Vygotsky, L. S. (1962). *Thought and language* (E. Hanfmann & G. Vakar, Eds.; 1st ed.). MIT Press. (Original work published 1934)

Vygotsky, L. S. (1963). *Thought and language*. MIT Press. (Original work published 1934)

Vygotsky, L. S. (1966). Development of the higher mental functions. In *Psychological research in the U.S.S.R.*, Aleksei Nikolaevich Leontyev, Aleksandr Romanovich Luria, and A. A. Smirnov (Eds.) (pp. 11–45). Progress.

Vygotsky, L. S. (1974). *The psychology of art*. MIT Press. (Original work published 1925)

Vygotsky, L. S. (1978). *Mind in society: The development of higher psychological processes*. Harvard University Press.

Vygotsky, L. S. (1986). *Thought and language*. Alex Kozulin (Ed.). 2nd ed. MIT Press. (Original work published 1934)

Vygotsky, L. S. (1989). Concrete human psychology. *Soviet Psychology, 27*(2), 53–77. (Original work written 1929)

Vygotsky, L. S. (1994a). The problem of the cultural development of the child. In R. V. der Veer & J. Valsiner (Eds.), *The Vygotsky reader* (pp. 57–72). Blackwell. (Original work published 1929)

Vygotsky, L. S. (1994b). The Socialist alteration of man. In R. van der Veer & J. Valsiner (Eds.), *The Vygotsky reader* (pp. 175–184). Blackwell. (Original work published 1930)

Vygotsky, L. S. (1997a). On psychological systems. In R. W. Rieber & J. Wollock (Eds.), *The collected works of L.S. Vygotsky: Vol. 3: Problems of the theory and history of psychology* (pp. 91–107). Springer. (Original work published 1930)

Vygotsky, L. S. (1997b). Psychology and the theory of the localization of mental functions. In R. W. Rieber & J. Wollock (Eds.), *The collected works of L.S. Vygotsky: Vol. 3: Problems of the Theory and History of Psychology* (pp. 139–144). Springer. (Original work published 1934)

Vygotsky, L. S. (1997c). *The collected works of L.S. Vygotsky: Vol. 4: The history of the development of higher mental functions*. Plenum. (Original work written 1931)

Vygotsky, L. S. (1997d). The instrumental method in psychology. In R. W. Rieber & J. Wollock (Eds.), *The collected works of L.S. Vygotsky: Vol. 3: Problems of the theory and history of psychology* (pp. 85–89). Springer. (Original work written 1930)

Vygotsky, L. S. (1997e). The methods of reflexological and psychological investigation. In R. W. Rieber & J. Wollock (Eds.), *The collected works of L.S. Vygotsky: Vol. 3: Problems of the theory and history of psychology* (pp. 35–50). Plenum.

Vygotsky, L. S. (1997f). The problem of consciousness. In Robert W. Rieber and Jeffrey Wollock (Ed.), *The collected works of L.S. Vygotsky: Vol. 3: Problems of the theory and history of psychology* (pp. 129–138). Plenum. (Original work written 1934)

Vygotsky, L. S. (2012). *Thought and language* (E. Hanfman, G. Vakar, & A. Kozulin, Eds.; 3rd ed.). MIT Press. (Original work published 1934)

Vygotsky, L. S., & Luria, A. R. (1993). *Studies on the history of behavior: Ape, primitive, and child*. Erlbaum. (Original work published 1930)

Vygotsky, L. S., & Luria, A. R. (1994). Tool and symbol in child development. In R. Van der Veer & J. Valsiner (Eds.), *The Vygotsky reader* (pp. 99–174). Blackwell. (Original work published 1930)

Wall, A., & Spinuzzi, C. (2018). The art of selling-without-selling: Understanding the genre ecologies of content marketing. *Technical Communication Quarterly, 27*(2), 137–160. https://doi.org/10.1080/10572252.2018.1425483

Weber, S. (2003). Boundary-crossing in the context of intercultural learning. In T. Tuomi-Gröhn & Y. Engeström (Eds.), *Between school and work: New perspectives on transfer and boundary-crossing* (pp. 157–178). Pergamon.

Wegerif, R. (2008). Dialogic or dialectic? The significance of ontological assumptions in research on educational dialogue. *British Educational Research Journal, 34*(3), 347–361. https://doi.org/10.1080/01411920701532228

Weinberg, E. (1974). *The development of sociology in the Soviet Union*. Routledge & Kegan Paul.

Wertsch, J. V. (Ed.). (1981). *The concept of activity in social psychology*. M.E. Sharpe.

Wertsch, J. V. (1985). *Vygotsky and the social formation of mind*. Harvard University Press.

Wertsch, J. V. (2007). Mediation. In H. Daniels, M. Cole, & J. V. Wertsch (Eds.), *The Cambridge companion to Vygotsky* (pp. 178–192). Cambridge.

Whyte, W. F. (1984). *Learning from the field: A guide from experience*. Sage.

Whyte, W. F. (2012). *Street corner society: The social structure of an Italian slum*. University of Chicago Press.

Whyte, W. F., & Hamilton, E. L. (1964). *Action research for management*. Irwin.

Winograd, T., & Flores, F. (1986). *Understanding computers and cognition: A new foundation for design*. Addison-Wesley.

Winsor, D. A. (1994). Invention and writing in technical work: Representing the object. *Written Communication, 11*(2), 227–250.

Winsor, D. A. (1996). *Writing like an engineer: A rhetorical education*. Erlbaum.

Winsor, D. A. (1998). Rhetorical practices in technical work. *Journal of Business and Technical Communication, 12*(3), 343–370.

Winsor, D. A. (1999). Genre and activity systems: The role of documentation in maintaining and changing engineering activity systems. *Written Communication, 16*(2), 200–224.

Winsor, D. A. (2000). Ordering work: Blue-collar literacy and the political nature of genre. *Written Communication, 17*(2), 155–184.

Winsor, D. A. (2001). Learning to do knowledge work in systems of distributed cognition. *Journal of Business and Technical Communication, 15*(1), 5–28.

Winsor, D. A. (2007). Using texts to manage continuity and change in an activity system. In M. Zachry & C. Thralls (Eds.), *Communicative practices in workplaces and the professions* (pp. 3–20). Baywood.

Witte, S. P. (2005). Research in activity: An analysis of speed bumps as mediational means. *Written Communication, 22*(2), 127–165. https://doi.org/10.1177/0741088305274781

Worthen, H. (2014). Projects, activity systems: Figuring out how to use these concepts. In A. Blunden (Ed.), *Collaborative projects: An interdisciplinary study* (pp. 356–360). Brill.

Wortis, J. (Ed.). (1950). On pedological distortions in the commissariats of education. In *Soviet psychiatry* (pp. 242–245). Williams & Wilkins.

Yamazumi, K. (2008). A hybrid activity system as educational innovation. *Journal of Educational Change, 9*(4), 365–373. https://doi.org/10.1007/s10833-008-9084-8

Yamazumi, K. (2009). Expansive agency in multi-activity collaboration. In A. Sannino, H. Daniels, & K. D. Gutierrez (Eds.), *Learning and expanding with activity theory* (pp. 212–227). Cambridge University Press.

Yamazumi, K. (2020). An interview with Yrjö Engeström and Annalisa Sannino on fourth-generation activity theory. *Actio: An International Journal of Human Activity Theory*, *9*(4), 87–98. https://doi.org/10.5565/rev/jtl3.709

Yasnitsky, A. (2016a). The archetype of Soviet psychology: From the Stalinism of the 1930s to the "Stalinist science" of our time. In A. Yasnitsky & R. Van der Veer (Eds.), *Revisionist revolution in Vygotsky studies* (pp. 3–26). Routledge.

Yasnitsky, A. (2016b). Unity in diversity: The Vygotsky-Luria circle as an informal personal network of scholars. In A. Yasnitsky & R. Van der Veer (Eds.), *Revisionist revolution in Vygotsky studies* (pp. 27–49). Routledge.

Yasnitsky, A. (2018a). *Vygotsky: An intellectual biography*. Routledge.

Yasnitsky, A. (2018b). Vygotsky's science of Superman: From Utopia to concrete psychology. In A. Yasnitsky (Ed.), *Questioning Vygotsky's legacy: Scientific psychology or heroic cult* (pp. 1–21). Routledge.

Yasnitsky, A. (2019). *Questioning Vygotsky's legacy*. New York: Routledge.

Yasnitsky, A. (2020a). *A history of Marxist psychology: The golden age of Soviet science*. Routledge.

Yasnitsky, A. (2020b). A new history of psychology: Soviet, Russian, Marxist. In A. Yasnitsky (Ed.), *A history of Marxist psychology: The golden age of Soviet science* (pp. 3–21). Routledge.

Yasnitsky, A., & Van der Veer, R. (2016). *Revisionist revolution in Vygotsky studies: The state of the art*. Routledge.

Yin, R. K. (2003). *Case study research: Design and methods* (3rd ed.). SAGE Publications.

Zaporozhets, A. V., Gal'prin, P. Ia., & El'konin, D. B. (1995). Problems in the psychology of activity. *Journal of Russian and East European Psychology*, *33*(4), 12–34. (Original work published 1969)

Zavershneva, E. (2016a). "The way to freedom": Vygotsky in 1932. In A. Yasnitsky & R. Van der Veer (Eds.), *Revisionist revolution in Vygotsky studies* (pp. 127–140). Routledge.

Zavershneva, E. (2016b). Vygotsky the unpublished: An overview of the personal archive (1912–1934). In A. Yasnitsky & R. Van der Veer (Eds.), *Revisionist revolution in Vygotsky studies* (pp. 94–126). Routledge.

Zavershneva, E. (2017). The problem of consciousness in Vygotsky's cultural-historical psychology. In A. Yasnitsky, R. van der Veer, & M. Ferrari (Eds.), *The Cambridge handbook of cultural-historical psychology* (pp. 63–100). Cambridge University Press.

Zavershneva, E., & Van Der Veer, R. (Eds.). (2018). *Vygotsky's notebooks: A selection*. Springer.

Zinchenko, V. P. (1985). Vygotsky's ideas about units for the analysis of mind. In J. V. Wertsch (Ed.), *Culture, communication and cognition: Vol. 35: Vygotskian perspectives* (pp. 94–118). Cambridge.

Zinchenko, V. P. (1995). Cultural–historical psychology and the psychological theory of activity: Retrospect and prospect. In J. V. Wertsch, P. del Rio, & A. Alvarez (Eds.), *Sociocultural studies of mind* (pp. 37–55). Cambridge University Press.

Zinchenko, V. P., Leont'ev, A. N., Lomov, B. F., & Luria, A. R. (1974). Parapsychology: fiction or fact? *Soviet Review, 15*(3), 74–91. https://doi.org/10.2753/RSS1061-1428150374

INDEX